Practical
ACOUSTICS

By

Stephen Kamichik

PROMPT© **Publications** is an imprint of Howard W. Sams & Company, A Bell Atlantic Company, 2647 Waterfront Parkway, E. Dr., Indianapolis, IN 46214-2041.

International Standard Book Number: 0-7906-1169-4
Library of Congress Catalog Card Number: 98-67779

Acquisitions Editor: Loretta Yates
Editor: Pat Brady
Assistant Editor: J.B. Hall
Typesetting: Pat Brady
Cover Design: Christina Smith, Brian Rees
Graphics: Jeremy Adams, Joe Kocha, Terry Varvel
Illustrations and Other Materials: Courtesy of the Author

PRINTED IN THE UNITED STATES OF AMERICA

9 8 7 6 5 4 3 2 1

Practical
ACOUSTICS

Contents

Chapter 5

Chapter 6

Chapter 7

Chapter15

Chapter 16

Chapter 17

Chapter 18

Chapter 19

Appendix

PREFACE

Most universities do not offer formal courses on acoustics, and there are very few books written on the subject. This book was written as a textbook for universities and technical schools. A course on acoustics can be offered by either the physics department or the engineering department of a university.

The hobbyist can teach himself if he reads and understands each chapter. Each theory chapter has questions at the end to test the reader's understanding of the material presented. The practicing engineer can also use this book as a reference text and as a refresher course.

This book is divided into two parts. *Part One* contains eleven chapters of theory. *Part Two* contains eight chapters full of interesting and exciting new projects.

Chapter 1 is an introduction to acoustics and it defines acoustics as the physics of sound.

Chapter 2 deals with simple harmonic motion and vibrations.

Chapter 3 discusses plane and spherical acoustic waves.

Chapter 4 introduces the decibel and how it relates to sound.

Chapter 5 deals with the transmission of sound through media.

Chapter 6 describes microphones and loudspeakers.

Chapter 7 explains the differences between noise, music and speech.

Chapter 8 discusses the anatomy of the human ear. It also explains how we hear.

Chapter 9 deals with architectural acoustics.

Chapter 10 discusses underwater acoustics.

Chapter 11 discusses ultrasonics and it defines ultrasonics as sound waves that occur at frequencies higher than the limits of human hearing.

Chapter 12 covers how to design and build world-class speaker systems. It is an in-depth look at speakers and speaker enclosures.

Chapter 13 describes a nine-band graphic equalizer. It can be used to correct for poor room acoustics or speaker system deficiencies.

Chapter 14 explains how to design active crossover networks.

Chapter 15 discusses the construction of a surround-sound home theater system.

Chapter 16 details the construction of an ultrasonic remote control.

Chapter 17 discusses the construction of an ultrasonic radar for automobiles. This all-new project illuminates a warning light when your vehicle is within about twelve inches of another vehicle.

Chapter 18 explains the LM381, LM382 and LM387 dual audio preamplifier integrated circuits. This chapter includes several practical circuits.

Chapter 19 discusses passive and active tone control circuits. This chapter also includes several practical circuits.

Part One: Acoustics

Chapter 1

Introduction

Acoustics is the physics of sound. Even though the basic theory of acoustics is about vibration and wave propagation, acoustics is a multidisciplinary science.

The physicist investigates the properties of matter by using concepts of wave propagation in different mediums. The acoustical engineer is concerned with the accurate reproduction of sound and with the conversion of electrical and mechanical energy into acoustical energy by using transducers. He is also interested in the design of acoustical transducers. The architect is interested in the isolation and absorption of sound in buildings. He also studies echo prevention and controlled reverberation in music halls and auditoriums. The musician is interested in how to obtain rhythmic combinations of tones using stringed musical instruments, air column musical instruments and percussion musical instruments.

Psychologists and physiologists study human speech and hearing mechanisms. They also study hearing phenomena, the reaction of people to sounds and music, and the psychoacoustic criteria for the comfort level of noise and pleasant listening conditions. Linguists are concerned with the subjective perception of complex noises and with the production of synthetic speech.

Sounds are created by sound *waves* traveling through a medium. The sound source causes air (or other medium) particles to vibrate; that is, to move up and down. The vibrating medium propagates the sound waves from the (sound) source to the (sound) receiver. Your eardrums pick up the air vibration and your inner ears convert the vibration into signals for your brain.

Sound waves oscillate at various frequencies. Low oscillations or frequencies generate deep bass sounds. High oscillations or frequencies generate high pitched sounds, such as breaking glass. Human beings can hear sounds in the range of 20 to 20,000 hertz. Dogs can hear sounds higher than 20,000 hertz and bats can detect sounds at 30,000 hertz.

Sound waves can be *reflected* waves or *direct* waves. An *echo* is a reflected sound. *Reverberation* is the algebraic sum of all echoes or reflected waves.

Ultrasonic sound waves are waves oscillating at more than 20,000 hertz. Ultrasonic sound waves are useful in oceanography, medicine and industry.

High levels of noise are generated by airplanes, heavy industry and household appliances. Ear damage, physical irritation and psychological trauma are caused by loud noises. It is therefore important that we understand the causes and effects of sound in order to control sound.

Problems

1-1.   Define acoustics.

1-2.   Define a transducer.

1-3.   What three things are required for sound?

1-4.   Does sound travel in a vacuum? Why?

1-5.   What range of frequencies can a human being hear?

1-6.   Define an echo.

1-7.   Define reverberation.

1-8.   Define ultrasonic sound.

Chapter 2

Sound Waves

Sound waves are generated by moving particles in an elastic medium. Sound requires a source, a medium and a receiver. The source must be vibrating within the frequency range of the receiver.

When the source moves forward from its static equilibrium position, it pushes or compresses the air ahead of it. Simultaneously, an empty space or rarefaction appears behind the sound source. Air rushes in to fill the empty space. The moving air becomes a sound source and the process is repeated over and over again. The compression of air is transferred to remote parts and the air is set into motion. The air motion generates sound waves. Sound is heard. The human ear detects the air disturbances and generates the auditory sensation of sound. Fluids and solids possess inertia and elasticity and can therefore transmit sound waves.

The propagation of sound waves involves the transfer of kinetic energy and potential energy through space. The kinetic energy component results in the movement of the particles of the medium. The potential energy component results in the elastic displacement of the particles of the medium.

Sound waves spread out in all directions around the sound source. Sound waves can be reflected, refracted, scattered, diffracted, interfered and absorbed. A medium is required for the propagation of sound waves. There is no sound transmission in a vacuum. The speed of sound depends on the density and temperature of the medium.

Transverse wave motion occurs when the medium disturbance is perpendicular to the direction of propagation of the wave as shown in *Figure 2-1*. A pendulum is an example of transverse vibration.

Longitudinal wave motion occurs when the medium disturbance is parallel to the direction of propagation of the wave as shown in *Figure 2-1*. A coil spring with a weight is an example of longitudinal vibration. Longitudinal vibration is generated by alternately compressing and stretching the spring. Sound waves are longitudinal waves.

SIMPLE HARMONIC MOTION

A particle in rectilinear motion has simple harmonic motion when its acceleration is proportional to its displacement from a fixed point. The direction of acceleration of the particle is towards the fixed point.

A particle has rectilinear motion if it is moving along a straight line. The velocity of a particle is its change in position, dx, per unit time, dt. Mathematically, $v = dx/dt$; velocity is the first derivative of displacement, x. The acceleration of a particle is its change in velocity, dv, per unit time, dt. Mathematically, $a = dv/dt$; acceleration is the first derivative of velocity, v. Acceleration is also the second derivative of displacement, x; that is, $a = d^2x/dt^2$.

Simple harmonic motion can be represented mathematically by $a = -w^2x$, where w is the circular frequency in rad/sec. The solution is $x(t) = 1$ A sin wt + B cos wt, where A and B are arbitrary constants. The solution may be rewritten as:

$$x(t) = SQRT(A^2 + B^2)\sin(wt + cc)$$

$$x(t) = SQRT(A^2 + B^2)\cos(wt - dd)$$

where: cc and dd are phase angles in radians.

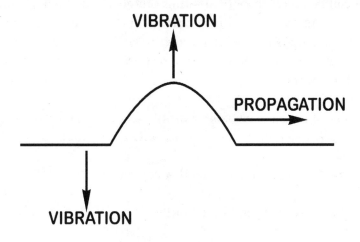

VIBRATION

PROPAGATION

VIBRATION

TRANSVERSE WAVE

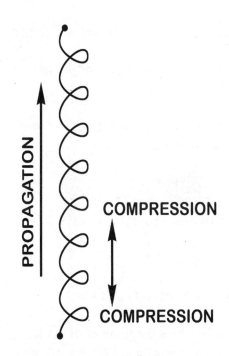

PROPAGATION

COMPRESSION

COMPRESSION

LONGITUDINAL WAVE

Figure 2-1.
Transverse and
longitudinal wave
propagation.

Simple harmonic motion can be either a sine or cosine function of time. It can be represented by rotating vectors as shown in *Figure 2-2*. Vector r has a constant magnitude and it is rotating counterclockwise at a constant angular velocity, w. The vector's projections on the x-axis and the y-axis are respectively cosine and sine functions of time.

A harmonic wave has a sinusoidal profile. A harmonic wave moving in the *positive* x direction with velocity c is written as:

$$u(x, t) = A \sin m(x - ct) \text{ and } A \cos m(x - ct)$$

where: A is the amplitude of the harmonic progressive waves.

A harmonic wave moving in the *negative* x direction with velocity c is written as:

$$u(x, t) = A \sin m(x + ct) \text{ and } A \cos m(x + ct)$$

where: A is the amplitude of the harmonic progressive waves.

A *spherical* wave moving away from the origin of the coordinate with velocity c is represented by:

$$u(r, t) = (A/r)f(ct - r)$$

A spherical harmonic progressive wave is represented by:

$$u(r, t) = \left(\frac{A}{r}\right)e^{j(wt-kr)}$$

where: $j = SQRT(-1)$ and $k = \frac{1}{L}$

The wave number, k, is the number of cycles of the wave per unit length. The wave repeats itself after a distance $L = 2PI/m$, where PI = 180 degrees of rotation. L is the wavelength which is the distance between a pair of corresponding points in two consecutive waves.

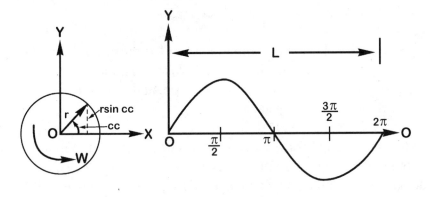

SINE FUNCTION

Figure 2-2.
Simple harmonic
motion.

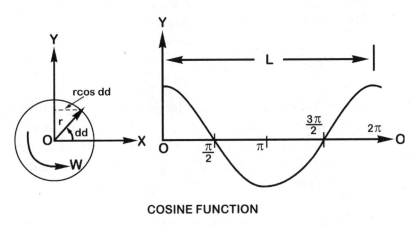

COSINE FUNCTION

VIBRATIONS

All systems that possess mass and elasticity are capable of relative motion. *Vibration* is any motion that repeats itself periodically. The *period* is the time required for one complete vibration. The *frequency* is the number of vibrations in one second.

A *transient* or *free vibration* is the periodic motion of a system when that system is displaced from its static equilibrium position. Friction causes the vibration to decrease with time. The result is an exponentially decaying vibration as shown in *Figure 2-3*. Mathematically, the exponentially decaying vibration is:

$$x(t) = e^{-dw1t}(A \sin w2t + B \cos w2t)$$

where: d is the damping factor, w1 is the natural circular frequency in rad/sec, w2 is the natural damped circular frequency in rad/sec, and A and B are arbitrary constants.

When an external force acts upon a system during its vibratory motion, the resulting motion is called *forced vibration*. The system tends to vibrate at its own natural frequency as well as to follow the frequency of the excitation force. When the system is damped, the vibration, at its own natural frequency, dies out and the system will only vibrate at the frequency of the external force. The resulting motion is called *steady-state vibration* or *response* of the system. Mathematically, it is written:

$$x(t) = \frac{F \cos(wt - cc)}{SQRT[(k - m^2)^2 + c^2]}$$

where: F is the magnitude of the external force, k is the elasticity of the system, m is the mass of the system, c is the damping coefficient, w is the frequency of the external force in rad/sec, and cc is the phase angle.

Resonance occurs when the frequency of the external force is the same as the natural frequency of the system. The amplitude of vibration increases and it is only restricted by damping forces in the system. If there are no damping forces in the system, the system can self-destruct.

The damping forces absorb energy and dissipate the energy as heat. The amplitude of free vibration decreases because the system is constantly losing energy. Energy must be continuously supplied to maintain steady-state vibration.

If there are no damping forces, the system continues to vibrate. The total energy is constant and it is equal to either the maximum potential energy or the maximum kinetic energy.

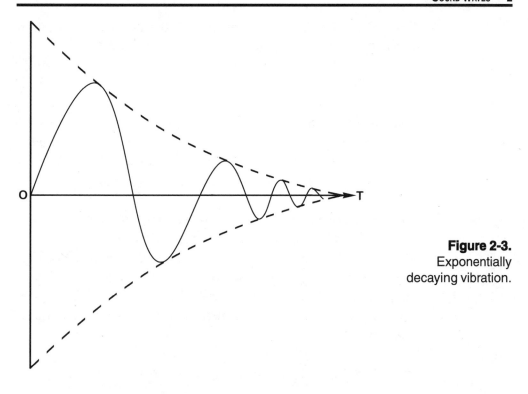

Figure 2-3.
Exponentially
decaying vibration.

A *string* is a vibrator with continuous media characteristics. It is the simplest medium of wave transmission. A string has its mass uniformly distributed along its length. A vibrating string generates *standing waves*. A standing wave is two waves of the same frequency and amplitude traveling in opposite directions through a medium. Mathematically:

$$y(x, t) = f1(x - at) + f2(x + at)$$

where: f1 and f2 are arbitrary functions.

A wave of arbitrary shape traveling in the positive x direction with velocity a is represented by:

$$f1(x - at)$$

A wave of arbitrary shape traveling in the negative x direction with velocity a is represented by:

f2(x + at)

A *bar* is a body that is elongated in one direction. It is made of homogeneous, isotropic material. It is free of transverse constraints. Mathematically, a bar may be thought of as several strings laminated together. A vibrating bar also generates standing waves.

A *membrane* is material of uniform thickness and it is held under homogeneous tension in a rigid frame. A membrane is flexible because its thickness is very small compared to its other two dimensions. When a membrane is excited, free vibration without damping takes place perpendicular to the surface of the membrane. A membrane generates two-dimensional standing waves. Mathematically:

y(x, z, t) = f1(mx + nz - at) + f2(mx + nz + at)

where: $m^2 + n^2 = 1$.

The two traveling waves have the same arbitrary profile and they are traveling in opposite directions along the x-axis and the z-axis with velocity a.

The thickness of a circular plate is not small compared to its other two dimensions. Mathematically, a circular plate may be thought of as several circular membranes laminated together. The general solution for free vibration of a circular plate is:

y(r, t) = [AJ(kr) + BI(kr)]e^{jwt}

where: A and B are arbitrary constants, J is the Bessel function of the first kind of order zero, j = SQRT(-1), and I is the Bessel hyperbolic function.

Problems

2-1.   How are sound waves generated?

2-2.   What is the kinetic energy component of sound transmission?

2-3.   What is the potential energy component of sound transmission?

2-4.   What determines the speed of sound?

2-5.   Define transverse wave motion.

2-6.   Define longitudinal wave motion.

2-7.   Define rectilinear motion.

2-8.   Define simple harmonic motion.

2-9.   Define velocity.

2-10.   Define acceleration.

2-11.   Define wavelength.

2-12.   Define a vibration.

2-13.   Define the period of a vibration.

2-14.   Define the frequency of a vibration.

2-15.   Define resonance.

2-16.   Name some types of vibrators.

2-17.   Define a standing wave.

Chapter 3

Plane and Spherical Acoustic Waves

Sound waves are generated when an elastic or fluid medium is disturbed. Sound waves are progressive longitudinal sinusoidal waves which travel through a three-dimensional space. Plane acoustic waves are one-dimensional free progressive waves. If the reader assumes that the pressure is constant in the y and z directions, plane acoustic waves travel in the x direction. The wavefronts are infinite planes perpendicular to the x-axis and they are parallel to each other at all times. The longitudinal wave motion of an infinite column of air enclosed in a smooth rigid tube with a constant cross-sectional area closely approximates plane acoustic wave motion.

The surface of a pulsating sphere exerts a force on the medium that surrounds it. A disturbance is produced uniformly in all directions, resulting in spherical acoustic waves. The spherical acoustic wave is mathematically one-dimensional because all points of the wave are related to the radial distance of the wavefront from the center of the pulsating sphere. Spherical acoustic waves do not change shape as they spread out. At great distances from the source, spherical acoustic waves resemble plane acoustic waves.

SPEED OF SOUND

The speed of sound is the speed of propagation of sound waves through the medium. It depends on the density and on the temperature of the medium. The speed of sound in air is:

$$c = SQRT(\frac{aP}{p}) \text{ m/sec}$$

where: a is the ratio of the specific heat of air at constant pressure to that at constant volume, P is the air pressure in newtons/m^2, and p is the density of air in kg/m^3.

At room temperature and standard atmospheric pressure, the speed of sound in air is 343 m/sec (1125 ft/sec), and increases 0.6 m/sec (2 ft/sec) for each degree centigrade increase of temperature. The speed of sound in air is independent of barometric pressure, frequency and wavelength. It is directly proportional to absolute temperature; that is:

$$\frac{c1}{c2} = SQRT(\frac{T1}{T2})$$

The speed of sound also depends on the elastic medium through which it is traveling, as listed in *Table 3-1*.

The speed of sound in solids with a large cross-sectional area is:

$$c = SQRT \frac{Y(1 - u)}{p(1 + u)(1 - 2u)} \text{ m/sec}$$

where: Y is Young's modulus of elasticity in nt/m^2, p is the density in kg/m^3, and u is Poisson's ratio.

When the cross-sectional area is small compared to the wavelength, the lateral effect in Poisson's ratio can be neglected and the speed of sound is:

$$c = SQRT(\frac{Y}{p}) \text{ m/sec}$$

The speed of sound in fluids is:

$$c = \text{SQRT} \left(\frac{B}{p} \right) \text{ m/sec}$$

where: B is the bulk modulus and p is the fluid density in kg/m^3.

When an aircraft travels at more than 742.5 miles per hour, it catches up with its own sound waves. A barrier of compressed molecules of air must be penetrated. This requires jet engines and a strongly constructed aircraft.

Medium	Velocity	
Gases	ft/sec	m/sec
Carbon dioxide (0° C)	846	258
Oxygen (0° C)	1041	317
Air (0° C)	1089	332
Hydrogen (0° C)	4165	1269.5
Liquids		
Water (9° C)	4708	1435
Seawater (9° C)	4756	1450
Solids		
Brass	11480	3500
Oak	12620	3850
Glass	16410	5000
Iron	16410	5000
Aluminum	16740	5104

Table 3-1.
Speed of sound in different medium.

PLANE ACOUSTIC WAVES

In the analysis of plane acoustic waves in a rigid tube with a zero viscosity medium, we assume a homogeneous and continuous fluid medium, an adiabatic process and an isotropic and elastic medium. There is motion of the fluid along the longitudinal axis of the tube when the fluid medium is disturbed. The small variations in pressure and density about the equilibrium state can be mathematically represented by the one-dimensional wave equation:

$$d^2u/dt^2 = c^2d^2t/dx^2$$

where: u is the instantaneous displacement.

The partial differential wave equation has the same form as those for free longitudinal vibrations of bars and for free transverse vibration of strings. The general solution for the wave equation is:

$$u(x,t) = f1(x-ct)f2(x+ct) = Ae^{j(wt-kx)} + Be^{j(wt+kx)}$$

where: k = w/c is the wave number, and A and B are arbitrary constants.

Plane acoustic waves have three important elements: particle displacement, acoustic pressure and density change or condensation.

Particle displacements from their equilibrium positions are amplitudes of motion of small constant volume elements of the fluid medium possessing average identical properties and is written:

$$u(x, t) = Ae^{j(wt-kx)} + Be^{j(wt+kx)} = A\cos(wt - kx) + B\cos(wt + kx)$$

Acoustic pressure is the total instantaneous pressure at a point minus the static pressure. Acoustic pressure is also known as the *excess* pressure. The effective sound pressure at a point is the root mean square value of the instantaneous sound pressure over one complete cycle at that point.

The acoustic pressure, P, is:

$$P = pc[B \sin(wt + kx) - A \sin(wt - kx)]$$

Density change is the difference between the instantaneous density and the constant equilibrium density of the medium at any point and it is written:

$$s = \frac{(p-p1)}{p1}$$

where: p1 is the constant equilibrium density.

When plane acoustic waves are traveling in the positive x direction, particle displacement lags particle velocity, condensation and acoustic pressure by ninety degrees. When plane acoustic waves are traveling in the negative x direction, acoustic pressure and condensation lag particle displacement by 90 degrees. Particle velocity leads condensation particle displacement by ninety degrees.

Acoustic intensity is the average power transmitted per unit area in the direction of wave propagation:

$$I = \frac{P^2_{rms}}{pc} \text{ watts/m}^2$$

where: P_{rms} is the root mean square pressure in nt/m^2, p is the density in kg/m^3, and c is the speed of sound in m/sec.

Sound energy density is the energy per unit volume in a given medium. Sound waves carry potential and kinetic energy. The instantaneous sound energy density is:

$$E_{ins} = \frac{pv^2 + Pv}{c} \text{ watt-sec/m}^3$$

where: p is the instantaneous density in kg/m^3, P is the static pressure nt/m^2, v is the particle velocity in m/sec, and c is the speed of sound in m/sec.

The *average* sound energy density is:

$$E_{av} = 0.5pv^2 \text{ watt-sec/m}^3$$

Specific acoustic impedance of a medium is the ratio of sound pressure to particle velocity:

$$z = \frac{P}{V} \text{ kg/m}^2\text{-sec or rayis}$$

If the harmonic; plane acoustic wave is traveling in the positive x direction:

$$z = pc \text{ rayis}$$

If the harmonic plane acoustic wave is traveling in the negative x direction:

$$z = -pc \text{ rayis}$$

The *specific acoustic impedance* varies from point to point in the x direction of standing waves, and is:

$$z = r + jx$$

where: r is the specific acoustic resistance, and x is the specific acoustic reactance.

SOUND MEASUREMENTS

The logarithmic scale is used to measure sound because of the very wide range of sound power, intensity and pressure. One *bel* is equal to ten decibels. The *decibel* is the lowest sound intensity that the human ear can detect.

Sound power level is $PWL = 10\log(W/W1)$ dB, where W1 is the reference power. The standard power reference is $W1 = 10^{-12}$ watt; therefore $PWL = (10\log W + 120)$ dB. A large rocket radiates an acoustical power of about 10^7 watts or 120 dB. A very soft whisper radiates an acoustical power of about 10^{-10} watts or 20 dB.

Sound *intensity* level is:

$$IL = 10 \log(I/I1) = (10 \log I + 120) \text{ dB}$$

where: $I1 = 10^{-12}$ watt/m^2 .

Sound *pressure* level is:

$$SPL = 20 \log(P/P1) = (20 \log P + 94) \text{ dB}$$

where: $P1 = 2 \times 10^{-5}$ nt/m^2 (or 0.0002 microbar).

For vibration measurements, the *velocity* level is:

$$VL = 20 \log \left(\frac{v}{v1}\right) \text{ dB}$$

where: the standard velocity reference v1 is 10^{-8} m/sec.

The *acceleration* level is:

$$AL = 20 \log \left(\frac{a}{a1}\right) \text{ dB}$$

where: the standard acceleration reference a1 is 10^{-5} m/sec^2.

RESONANCE OF AIR COLUMNS

Acoustic resonance of air columns is a tuned response where the receiver is excited to vibrate by sound waves having the same frequency as its natural frequency. Resonant response depends on the distance between the sound source and the receiver and on the coupling of the medium between them.

The *Helmholtz resonator* is a spherical container filled with air. It has a large opening at one end and a much smaller opening at the other end. The Helmholtz resonator detects the frequency of vibration to which it is tuned. It uses the principle of air column resonance. Sound enters the large opening which is amplified at the small opening.

Half wavelength resonance of air columns is observed when the phase change on reflection is the same at both ends of the tube; that is, either two nodes or two antinodes. The effective lengths of air column and its resonant frequencies are:

$$EL = \frac{Li}{2}, f = \frac{c}{L} = \frac{ic}{2L}, i = 1, 2...$$

where: L is the wavelength and c is the speed of sound.

Quarter wavelength resonance of air columns occurs when there is no change in phase at one end of a stationary wave with a 180 degree phase change at the other end. The effective lengths of air column and its resonant frequencies are:

$$EL = \frac{L(2i-1)}{4}, f = \frac{c(2i-1)}{4L}, i = 1, 2, 3,...$$

Usually, an open end of a tube is an *antinode* and a closed end of a tube is a *node*.

DOPPLER EFFECT

The Doppler effect is the apparent change in frequency when there is motion between the sound source and the listener. The sound frequency increases when the sound source approaches the listener. The sound frequency decreases when the sound source moves away from the listener. The sound source, the listener, or both may be in motion for the Doppler effect to occur.

The frequency of a sound depends on the number of sound waves reaching the ear each second, and is:

$$f' = f(c-v)/(c-u) \text{ Hz}$$

where: f' is the observed frequency, c is the speed of sound, v is the velocity of the observer relative to the medium, f is the frequency of the sound source, and u is the velocity of the sound source.

SPHERICAL ACOUSTIC WAVES

A pulsating sphere generates spherical acoustic waves. The medium is three-dimensional. The spherical wavefront is one-dimensional because all points on the wave are related to the radial distance of the wavefront from the center of the sphere.

The three-dimensional wave equation in spherical coordinates is:

$$\frac{d^2(rP)}{dt^2} = \frac{c^2 d^2(rP)}{dr^2}$$

> where: r is the radial distance of the source from the wavefront, P is the acoustic pressure, and c is the speed of sound.

The solution for the wave equation is:

$$P(r,t) = \frac{f(ct-r)}{r} + \frac{g(ct+r)}{r}$$

> where: f and g are arbitrary functions.

For harmonic progressive spherical acoustic waves, the *particle displacement* is:

$$u = \frac{-P(1/r + jk)}{w^2 p}$$

The *particle velocity* is:

$$v = \frac{p(1/r + jk)}{jwp}$$

The *condensation* is:

$$s = \frac{P}{pc^2}$$

> where: P is the acoustic pressure, p is the medium density.

Acoustic intensity is the average rate of flow of sound energy per unit area. For spherical acoustic waves:

$$I = \frac{0.5P^2}{\rho c} \text{ watts/m}^2$$

Energy density of a spherical acoustic wave at any instant is the sum of kinetic energy and potential energy per unit volume.

$$E = 0.25(\frac{\rho v^2 + P^2}{\rho c^2}) \text{ joules/m}^3$$

Specific acoustic impedance is the ratio of acoustic pressure to the velocity at any point in the wave. For harmonic progressive spherical acoustic waves:

$$z = \rho ckr (\frac{kr}{1+k^2r^2} + \frac{j}{1+k^2r^2}) \text{ rayis}$$

The real part is the specific acoustic resistance and the imaginary part is the specific acoustic reactance. The wave number is:

$$k = \frac{w}{c}$$

RADIATION OF SOUND

A sound source is an isotropic radiator if the sound waves radiated outward are symmetric and uniform in all directions. A pulsating sphere is the simplest isotropic radiator. When the dimensions of a sound radiator are small compared to the wavelength of the sound radiated, the radiator can be approximated by a pulsating sphere.

A large flat surface sound radiator, such as a membrane, is not an isotropic sound radiator. However, the radiation produced at any point by a large flat surface can be assumed to be equal to the sum of the radiation produced by an array of isotropic sound radiators.

Sound waves produced by most sources are directional for several reasons: the size and shape of the sound source, the radiation impedance, the mode of vibration of the surface of the radiator and the reaction of the fluid medium on the surface of the sound radiator. An *infinite baffle* is a large rigid surface. The presence of an infinite baffle near the sound source confines the sound radiation to one side of the surface. An infinite baffle also affects the directivity of the sound source.

The *directivity pattern* of a sound source is a graphical representation of the response of a sound radiator as a function of the direction of the transmitted sound waves in a specified plane for a specified frequency.

The *directivity ratio* is the ratio of the intensity at any point on the axis of the sound source to the intensity that would be produced at the same point by a simple source of equal strength:

$$D = \frac{I}{I1}$$

The *directivity index* or *gain* is the decibel expression for the directivity ratio:

$$d = 10\log D \text{ dB}$$

Beam width is the angle at which the sound intensity drops down to one-half of its value at the axial direction of the source.

Source strength is the product of the surface area and the velocity amplitude of a pulsating sphere:

$$Q = 4 \cdot PI \cdot vr^2$$

A hemispherical source mounted in an infinite baffle has one-half the strength of a similar spherical source having the same radius and velocity amplitude.

Acoustic doublet is an arrangement of two simple sound sources of identical strength and frequency. The directivity pattern of this array of sound sources depends on the distance between the two sources and on the phase between them.

Radiation impedance is the ratio of the force exerted by the sound radiator on the medium to the velocity of the radiator:

$$z = \frac{f}{v} \text{ kg/sec}$$

The *total impedance* acting on the radiator is the sum of its mechanical impedance and its radiation impedance:

$$z1 = R + j(wm - \frac{k}{w})$$

where: w is the frequency.

These impedances are functions of the frequency. The *velocity amplitude* changes as the frequency is varied:

$$v = \frac{f}{(z1 + z)}$$

Problems

3-1.   Where do spherical acoustic waves resemble plane acoustic waves?

3-2.   What factors affect the speed of sound?

3-3.   Calculate the speed of sound in air at 20 degrees Celsius.

   (Use: a = 1.4, P = 1.01×10^5 nt/m², and p = 1.21 kg/m³)

3-4.   Calculate the speed of sound in water where:

$B = 2.1 \times 10^9$ nt/m^2, and p = 998 kg/m^3.

3-5.   Calculate the speed of sound in copper where:

$Y = 12.2 \times 10^{10}$ nt/m^2, and p = 8900 kg/m^3.

3-6.   Name three important elements of plane acoustic waves.

3-7.   Define acoustic pressure.

3-8.   Define acoustic intensity.

3-9.   Calculate the acoustic intensity if:

$P_{rms} = 0.00002$ nt/m^2, p = 1.21 kg/m^3, and c = 343 m/sec.

3-10.   Define sound energy density.

3-11.   Define specific acoustic impedance.

3-12.   Calculate the specific acoustic impedance in air if:

particle velocity = 340 m/sec, and $P = 1.01 \times 10^5$ nt/m^2.

3-13.   Calculate the change in the sound power level when a speaker's output is increased from 5 watts to 50 watts.

3-14.   Define Doppler effect.

3-15.   An automobile emits a 100 Hz sound. The car is moving away from a stationary observer at a velocity of 10 m/sec. What frequency will the observer hear? If the car is moving towards the observer, what frequency will he hear?

3-16.   Define acoustic intensity.

3-17. Define the energy density of a spherical acoustic wave.

3-18. Define specific acoustic impedance.

3-19. What is an infinite baffle?

3-20. Define radiation impedance.

Chapter 4

Sound and the Decibel

In 1933, scientists H. Fletcher and W.A. Munson published the results of their research in sound. The Fletcher-Munson curves are shown in *Figure 4-1*.

The human ear is nonlinear in its response to sound in the range of 20 hertz to 20,000 hertz. The human ear has a *logarithmic response* to sound. The response of the human ear to sound waves is approximately proportional to the logarithm of the energy of the sound wave—it is not proportional to the energy of the sound wave.

The ear is most sensitive to sounds in the 2000 to 4000 hertz range. It requires 38 dB more power at 100 hertz for the threshold level of hearing than the power required at 1000 hertz for the threshold level of hearing.

As the sound level increases, the Fletcher-Munson curves tend to flatten out. The human ear is less sensitive to low and high frequencies than it is to medium frequencies. Some typical sound intensity levels are listed in *Table 4-1*.

The *volume control* of an amplifier raises or lowers the gain of the amplifier at all frequencies; that is, the frequency response is flat. The *loudness control* boosts low and high frequencies as well as attenuates medium frequencies at low volume settings. At higher volume settings, the loudness control, like the volume control, has a flat frequency response. The loudness control yields more realistic sound at all volume settings.

The ear is nonlinear in its response to sound. The logarithmic scale is required for sound measurements because the human ear has a logarithmic response to sound. The logarithmic scale converts a nonlinear scale into a linear scale. The human ear's response to the magnitude of sound is roughly proportional to the logarithm of the actual sound energy. The change in gain of an amplifier, expressed in *decibels* (dB), is a much better index of the effect of the sound on the human ear than if the change in gain of an amplifier is expressed as a voltage ratio or as a power ratio.

The *bel*, in honor of Alexander Graham Bell, is used to measure sound intensity. The bel is the logarithm to base 10 of the ratio of two powers:

$$\text{\# of bels} = \log_{10} \frac{P2}{P1}$$

where: P2 and P1 are the two powers being compared.

The bel is a cumbersome unit. A convenient unit is the *decibel*, which is one-tenth of a bel:

$$\text{\# of decibels} = dB = 10 \log_{10} \frac{P2}{P1}$$

The decibel is the lowest sound intensity that the human ear can detect. The decibel is a relative unit; that is, the decibel value must be compared to a reference level. For sound intensity, a reference level of one decibel corresponds to an acoustical field strength of 10^{-16} Watt/cm^2, the threshold level that the human ear can detect at 600 Hz. The threshold of pain is 130 dB. The human ear and brain, therefore, have a dynamic range of 130 dB, which is a ratio of ten trillion to one.

Decibel readings are obtained with a sound level meter. A microphone transforms sound energy into electrical energy. The electrical signal is amplified for measurements. The sound level meter is used to measure sound energy in decibels.

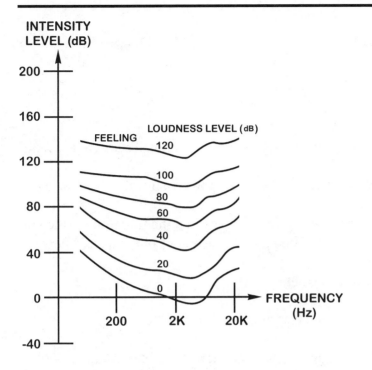

Figure 4-1.
Fletcher-Munson curves.

Type of Sound	Intensity Level (dB above 10^{-16} W/cm²)
Threshold of pain	130
Airplane	120
Subway	100
Niagara Falls	90
Conversational speech	70
Office	55
Residence	40
Whisper	20
Barely audible sound	0

Table 4-1.
Typical sound intensity levels.

Problems

4-1.   What type of response does the human ear have to sound?

4-2.   What is the frequency response range of the human ear?

4-3.   What is the difference between the functions of the volume control and the loudness control of an amplifier?

4-4.   Define the bel.

4-5.   How many bels are there if the power ratio is 100?

4-6.   How many decibels are there if the power ratio is 100?

4-7.   What is the threshold of pain with regard to sound intensity?

Chapter 5

Transmission of Sound

Sound waves traveling through an elastic medium may be reflected, refracted, diffracted, scattered, interfered or absorbed. Incident, reflected and transmitted sound waves at a plane interference are shown in *Figure 5-1*.

Sound waves are *reflected* when they are turned or thrown back. Sound waves are *refracted* when they change direction because they are passing from one elastic medium to another elastic medium. Sound waves are *diffracted* when they change direction because they are passing through an opening or because they are blocked by an obstacle. Sound waves are *scattered* when they separate and go in different directions. Sound waves may be *interfered* or impeded. Sound waves may be *absorbed* by molecular action.

The transmission of sound waves involves the transfer of acoustic energy through the elastic medium through which the sound waves are traveling.

SOUND TRANSMISSION THROUGH MEDIA

Sound power *reflection coefficient* is the ratio of the reflected flow of sound energy to the incident flow of sound energy. This definition applies to a sinusoidal plane acoustic wave traveling from one fluid medium to another fluid medium at normal incidence along the plane interference of the two media.

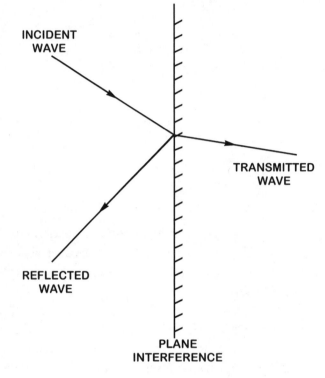

Figure 5-1.
Incident, reflected and
transmitted sound waves.

$$a_r = (\frac{p2c2 - p1c1}{p2c2 + p1c1})^2$$

where: p is the medium density, and c is the speed of sound
in that medium. The number (1 or 2) denotes the
medium.

Sound power *transmission coefficient* is the ratio of the transmitted sound power to the incident sound power:

$$a_t = \frac{4p1c1p2c2}{(p1c1+p2c2)^2}$$

Normal specific impedance characterizes the behavior of solids
with sound waves, when the sound waves are at normal or perpendicular incidence at the surface of the solid:

$$z_n = r_n + jx_n$$

where: r_n is the resistive component, and x_n is the reactive component.

$$a_r = (r_n - p1c1)^2 + \frac{x_n^2}{(r_n + p1c1)^2} + x_n^2 \text{ at normal incidence and}$$

$$a_t = \frac{4p1c1r_n}{(r_n + p1c1)^2} + x_n^2 \text{ at normal incidence.}$$

When sound waves are traveling at oblique (not perpendicular) incidence, the sound power reflection coefficient and the sound power transmission coefficient are:

$$a_r = \left(\frac{p2c2 \cos Ai - p1c1 \cos At}{p2c2 \cos Ai + p1c1 \cos At}\right)^2$$

where: Ai is the angle of incidence and At is the angle of refraction.

$$a_t = \frac{4p1c1p2c2 \cos Ai \cos At}{(p2c2 \cos Ai + p1c1 \cos At)^2}$$

Sound power reflection coefficient and sound power transmission coefficient at oblique incidence to the surface of a normally reacting solid are:

$$a_r = \frac{(r_n \cos Ai - p1c1)^2 + x_n^2 \cos^2 Ai}{(r_n \cos Ai + p1c1)^2 + x_n^2 \cos^2 Ai}$$

$$a_t = \frac{4p1c1r_n \cos Ai}{(r_n \cos Ai + p1c1)^2 + x_n^2 \cos^2 Ai}$$

The transmission of a sinusoidal plane acoustic wave from a fluid medium, through a second medium and into a third medium is similar to the transmission of sound waves through two media.

Part of the incident sound waves are reflected at the plane interference of the fluid media and part of the incident sound waves are transmitted through the boundaries. The sound power transmission coefficient for sound waves traveling from the first medium through the second medium and into the third medium is:

$$a_t = \frac{4p1c1p3c3}{(p1c1 + p3c3)^2 \cos^2 k_2 L + (\frac{p2c2 + p1c1p3c3}{p2c2})^2 \sin^2 k_2 L}$$

where: k_2 = w/c and is the wave number of medium 2, and L is the thickness of medium 2.

Transmission loss is the difference between the sound energy striking the surface which separates two spaces and the energy transmitted. It is expressed in decibels. Transmission loss cannot be measured but it is computed from sound pressure level measurements on both sides of the partition:

$$TL = 10 \log (\frac{Ii}{It}) \, dB$$

where: Ii is the incident sound intensity, and It is the transmitted sound intensity.

REFLECTION OF SOUND WAVES

Sound waves are reflected whenever there is a discontinuity and interference of two media in which the sound waves are propagated. The reflected sound wave depends on the incident wave, the angle of incidence, the reflecting surface and the characteristic impedance of the media. The reflected flow of sound energy is proportional to the square of the amplitude of the reflected sound wave.

A sound wave is at *normal incidence* to a surface when the wave is perpendicular to that surface, as shown in *Figure 5-2*. A sound wave is at *oblique incidence* to a surface when the wave is not perpendicular to that surface, as shown in *Figure 5-3*.

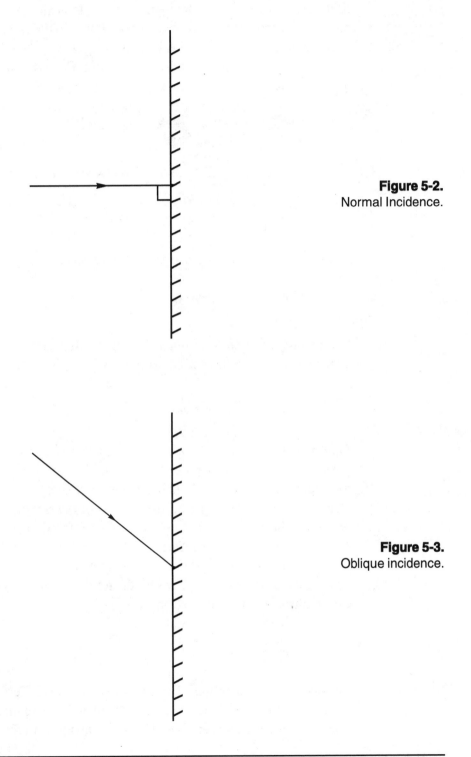

Figure 5-2.
Normal Incidence.

Figure 5-3.
Oblique incidence.

Standing wave ratio (SWR) is the ratio of acoustic pressure at an antinode to acoustic pressure at a node. It is also the ratio of maximum to minimum amplitudes in a standing wave. The SWR is an indication of the amount of sound energy reflected at the boundary:

$$\text{SWR} = \frac{P_{MAX}}{P_{MIN}} = \frac{A_{MAX}}{A_{MIN}} = \frac{P_i + P_r}{P_i - P_r}$$

If a sound wave is totally *reflected*, SWR is infinity, or:

$$\frac{P_r}{P_i} = 1$$

If a sound wave is totally *absorbed* (no reflection):

$$\text{SWR} = 1 \quad \text{or} \quad \frac{P_r}{P_i} = 0$$

Snell's law, or the *law of reflection*, states that the angle of incidence equals the angle of reflection. Snell's law is mathematically:

$$\frac{c_{before}}{(\sin A)_{before}} = \frac{c_{after}}{(\sin A)_{after}}$$

Echo is a delayed sound and it is the result of reflected sound. If the reflected sound occurs less than one-tenth second after the original sound, the human ear and brain cannot detect it. These sound reflections are called *reverberation* echo or *overlapping* echo. A *musical* echo is the rapid and successive reflection of a sound. *Flutter* echoes are pulses reflecting back and forth, with diminishing amplitude, from one end to the other end of an enclosure. Echoes are used in navigation, in the detection of submerged ships and in ultrasonic flaw detection.

REFRACTION OF SOUND WAVES

Refraction is the bending of sound waves because their velocity changes when the temperature and pressure of the medium changes. The transmitted waves are bent toward or away from the normal to the boundary in accordance with the speeds of sound in the media. *Transmitted waves* are the waves that travel through the boundary, as shown in *Figure 5-1*. The direction of the transmitted waves are not the same as that of the incident waves. If the angle of incidence is greater than the critical angle, all of the sound waves are reflected. Refraction of sound waves takes place in a single medium, such as the earth's atmosphere, or in a large body of fluid, such as the ocean. Refraction in the atmosphere occurs because the wind and temperature vary from place to place. Refraction in the ocean occurs because the temperature varies from place to place.

DIFFRACTION OF SOUND WAVES

Sound waves are diffracted because they spread around the edges of an obstacle. Sound waves are diffracted rather than reflected when the sound wavelength is approximately the same as the dimensions of the obstacle.

SCATTERING OF SOUND WAVES

Sound waves are scattered when the sound wavelength is large compared to the dimensions of the obstacle. The amplitude of the scattered waves far from the obstacle is directly proportional to the volume of the obstacle and inversely proportional to the square of the wavelength. Therefore, sounds of long wavelength scatter little. Sounds of short wavelength scatter a lot.

Diffuse echo is produced by the scattering of sound by several small obstacles. A *harmonic echo* is the differential scattering of a complex sound or noise of different frequencies.

INTERFERENCE OF SOUND WAVES

When sound waves of the same frequency and amplitude are superimposed, they either cancel or reinforce each other. The result at each point in the medium is the algebraic sum of the amplitudes of the two waves. *Destructive interference* occurs where the sound waves meet in opposite phase. *Constructive interference* occurs where the sound waves meet in phase.

Standing or *stationary waves* are formed from the interference of two sound waves of equal amplitude and frequency propagated through a medium along the same line and in opposite directions. There are nodes (zero amplitude) and there are antinodes (maximum amplitude). The medium is set into steady-state vibration. *Beats* are produced from the interference of two sound waves of slightly different frequencies.

FILTRATION OF SOUND WAVES

Filtration of sound is the elimination of some of the sound waves of definite frequencies and wavelengths. The rest of the sound waves are allowed to pass. *Acoustic filters*, such as mufflers and resonators, separate components of noise or sound on the basis of their frequency. They permit components of sound of one or more frequency bands to pass unattenuated. Acoustic filters attenuate components of sound in other frequency bands. A low-pass acoustic filter, a high-pass acoustic filter and a notch acoustic filter are illustrated in *Figure 5-4*.

ABSORPTION OF SOUND WAVES

A sound wave can lose some of its energy due to absorption. Viscous loss of sound energy in a fluid medium is caused by shear stresses set up in the medium by the passage of compressive waves through the medium. *Conduction* or heat losses are caused by the flow of heat from the warmer more compressed part of the medium to the cooler more expanded part of the medium. Molecular energy losses are due to thermal relaxation which results in exchanges of energy between different internal thermal states of the molecules. Sound energy is absorbed if the phase of these exchanges of energy differs from the phase of the sound waves.

Sound energy is absorbed in air more rapidly at higher frequencies and less rapidly at lower frequencies. Sound energy is absorbed in water due to the scattering effect of the nonhomogeneties in the structure of water. Sound energy is absorbed in solids due to the interactions between sound waves and electron motion, ferromagnetic effects and ferroelectric effects.

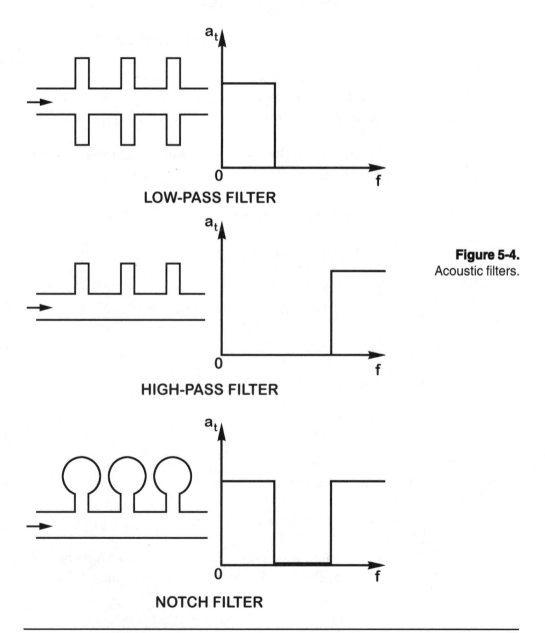

Figure 5-4.
Acoustic filters.

Problems

5-1. Define sound power reflection coefficient.

5-2. Define sound power transmission coefficient.

5-3. A beam of sound waves is normally incident on a plane interface of air and an infinite body of fluid. If half the sound energy is reflected, find the impedance of the fluid.

Where: In air, the speed of sound is 343 m/sec and the density of air is 1.21 nt/m^2.

5-4. Define transmission loss.

5-5. What is the transmission loss:

Where: the incident sound intensity is 10 W/m^2 and the transmitted sound intensity is 5 W/m^2 ?

5-6. Define SWR.

5-7. Prove that:

$$\frac{P_r}{P_i} = \frac{SWR - 1}{SWR + 1}$$

5-8. State Snell's law.

5-9. Define the critical angle.

5-10. If the sound wavelength is about the same dimension as the obstacle, what happens to the sound wave?

5-11.   If the sound wavelength is large compared to the dimension of the obstacle, what happens to the sound wave?

5-12.   What is destructive interference?

5-13.   What is constructive interference?

Chapter 6

Microphones and Loudspeakers

The microphone and the loudspeaker are *electroacoustic transducers*. A microphone converts acoustical energy into electrical energy. A loudspeaker converts electrical energy into acoustical energy. Microphones are used to record sound and to make acoustical measurements. Loudspeakers are used to reproduce and amplify sound.

MICROPHONES

Microphones are dynamic air pressure transducers. The moving coil, the velocity-ribbon and the magnetostriction microphones are *constant-velocity* microphones. The carbon, the condenser and the crystal microphones are *constant-amplitude* microphones. Sound pressure moves the diaphragm of the microphone.

TYPES OF MICROPHONES

The pressure-operated microphone, shown in *Figure 6-1*, uses the cyclic variation of air pressure when an elastic medium is vibrated. The pressure inside the microphone is kept at atmospheric level because there is a small hole in the microphone housing. The force acting on the diaphragm is proportional to the sound pressure and it is independent of frequency.

Figure 6-1.
Pressure-operated
microphone.

PRESSURE

MICROPHONE
HOUSING

DIAPHRAGM

HOLE

The pressure-gradient microphone, shown in *Figure 6-2*, has both surfaces of the diaphragm exposed to the sound pressure. The pressure-gradient microphone experiences a phase difference in sound pressure. The pressure difference, or *gradient*, causes the diaphragm to move and to produce a force that is proportional to frequency and to the path length d. A pressure-gradient microphone discriminates against sounds arriving at an angle to the axis of the microphone.

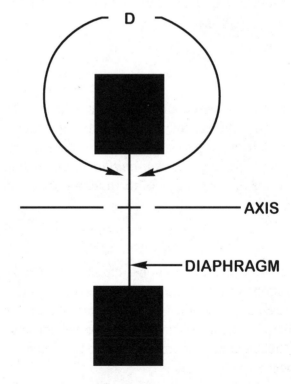

Figure 6-2.
Pressure-gradient
microphone.

D

AXIS

DIAPHRAGM

MICROPHONE SPECIFICATIONS

Sensitivity or open-circuit voltage response of a microphone is the voltage output for a sound pressure input of one microbar. A carbon microphone response is:

$$20 \log(\frac{M}{10}) \text{ dB}$$

where: M = EhA/Rs, E is the battery voltage, h is a resistance constant in ohms/m, A is the area of the diaphragm, R is the internal impedance of the microphone, s is the effective stiffness of the diaphragm in nt/m, and M is the microphone sensitivity.

Directivity or directional response characteristics of a microphone is the variation of the microphone output with different angles of incidence. It is usually represented by a polar graph or directivity characteristic as shown in *Figure 6-3*. The *directional response characteristics* of a unidirectional or cardioid microphone is the combination of the response characteristics of an omnidirectional and a bidirectional microphone. It discriminates against sounds from its sides and back. It receives all sounds from its front. Other unidirectional response characteristics may be obtained by the combination of different sizes of omnidirectional and bidirectional response characteristics.

Directional efficiency of a microphone is the ratio of the energy output due to simultaneous sounds at all angles to the energy output which would be obtained from an omnidirectional microphone with the same axial sensitivity.

RESONANCE

The effects of resonance on the performance of a microphone may be controlled. It can be made negligible by resistance control, by mass control and by compliance control.

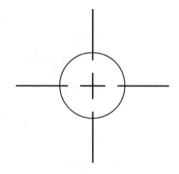

OMNIDIRECTIONAL

Figure 6-3.
Bidirectional response characteristics of a microphone.

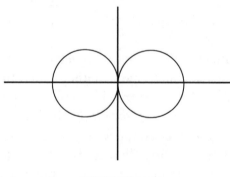

BIDIRECTIONAL

Resistance control is damping the diaphragm in order to reduce the amplitude of the vibration of the diaphragm. *Mass control* is lowering the resonant frequency below the working frequency by making the microphone housing thick and heavy. *Compliance control* is raising the resonant frequency above the working frequency.

CALIBRATING AND TESTING MICROPHONES

Microphones can be calibrated several ways. Direct known sound source, comparison, Rayleigh disc, radiometer, hot-wire microphone, motion of suspended particles, and the reciprocity technique are methods for calibrating microphones. Microphone calibration can be carried out either in an anechoic chamber or in a closed or reverberation chamber.

In an anechoic chamber the sound waves are purely progressive in a free field. In a closed chamber the acoustic intensity and energy are constant throughout.

A microphone can be tested for distortion and frequency response as shown in *Figure 6-4*. A sinewave signal is fed to a speaker which is located near the microphone. The microphone should be shielded from all sound sources except the loudspeaker. The microphone output is fed to an oscilloscope. Examine the oscilloscope pattern for microphone distortion. Vary the frequency of the sinewave signal to determine the frequency response of the microphone by viewing the oscilloscope pattern.

The choice of a microphone is determined by environmental conditions and by frequency response. Microphones should have high sensitivity, suitable directivity, uniform frequency response, minimum phase distortion and very little inherent or external noise.

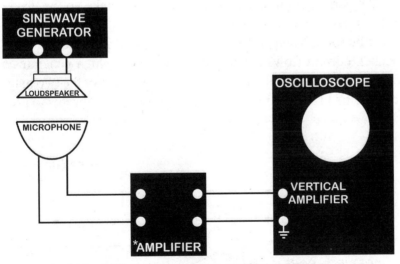

* AMPLIFIER MAY BE OMITTED IF OSCILLOSCOPE
VERTICAL AMPLIFIER GAIN IS SUFFICENT.

Figure 6-4.
Testing a microphone.

LOUDSPEAKERS

The loudspeaker converts electrical energy into acoustical energy. An electrical signal is applied to the speaker terminals. The speaker cone moves forward and backward in response to the electrical signal. The air around the speaker is pressurized and depressurized, producing sound waves. High frequency signals cause the speaker cone to vibrate quickly. Low frequency signals cause the speaker cone to vibrate slowly. Speakers should be efficient, should be able to handle high power, should have a flat frequency response and should have minimum distortion.

TYPES OF LOUDSPEAKERS

The most common loudspeaker is the *dynamic* speaker. The construction of a dynamic loudspeaker is shown in *Figure 6-5*. It has a voice coil which is immersed in a fixed magnetic field, F1. A powerful permanent magnet generates the fixed magnetic field, F1. The permanent magnet and the voice coil make up the *driver* of the speaker. The voice coil has many turns of fine wire wound on the bobbin. When an electrical audio signal is fed to the loudspeaker, current flows through the voice coil, which generates a second varying magnetic field, F2. The interaction of the two magnetic fields produces motion and the diaphragm vibrates to pro-

Figure 6-5.
Dynamic
loudspeaker
construction.

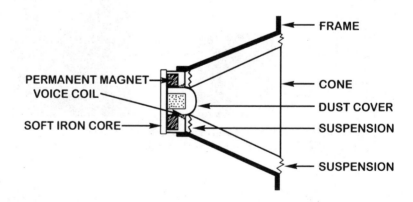

duce sound waves. The voice coil bobbin is attached to the speaker cone. When the coil moves in response to an electrical signal, the bobbin moves the cone and causes it to vibrate. The dust cap forms the center of the cone and keeps dust and debris from entering into the small gap between the voice coil and the permanent magnet core.

The speaker suspension must be flexible because it must allow the speaker cone to vibrate. The suspension attaches the speaker cone to the speaker frame. The dynamic loudspeaker has a low AC resistance or impedance, in the range of four ohms to 16 ohms. The DC resistance of a dynamic loudspeaker is about six ohms.

The *electrodynamic* loudspeaker operates in a similar manner to the dynamic speaker. Like the dynamic speaker, the electrodynamic speaker is sensitive to current. Unlike the dynamic loudspeaker, the magnetic field of an electrodynamic loudspeaker is energized by an external power source.

The *condenser* or electrostatic loudspeaker is sensitive to voltage. It therefore has a high impedance. It converts electrical audio signals into mechanical movements of the diaphragm. The diaphragm vibrations are caused by electrostatic forces of attraction and repulsion which are generated at the electrodes. The electrodes are energized by voltage to produce variation in capacitance. The electrodes have to be closely spaced; therefore, the electrostatic loudspeaker is not suitable for reproducing low frequency audio signals.

The *crystal* or piezoelectric loudspeaker operates on the theory that crystals expand or contract—that is, vibrate—when an alternating electric current is applied to the surfaces of the crystal. The crystal loudspeaker has a very limited low frequency response and a low power output. The piezoelectric loudspeaker makes an excellent tweeter or high frequency speaker. The piezoelectric effect was discovered by Jacque and Pierre Curie, who were French physicists.

LOUDSPEAKER SPECIFICATIONS

Acoustical power output of a speaker is:

$$W = \frac{(BL)^2 \, RI^2}{Z_m^2} \quad \text{or} \quad W = \frac{(BL)^2 \, RE^2}{Z_m^2 Z_i^2}$$

> where: B is the magnetic flux density, L is the length of the voice coil, R is the radiation resistance, I is the current, E is the applied voltage, Z_m is the total mechanical resistance and Z_i is the total input electrical impedance.

In a multispeaker system, the speakers must be matched in efficiency to produce an overall smooth frequency response. The speaker frequency response ranges must overlap to ensure that there are no holes in the frequency response curve.

The *frequency response* of a speaker is the range of frequencies that the speaker can reproduce. The *woofer* is a speaker designed to respond to frequencies less than about 3000 hertz. The *tweeter* is a speaker designed to respond to frequencies higher than about 4000 hertz. The *midrange* speaker is designed to respond to frequencies between one 1000 and 10,000 hertz.

Free-air resonance is the frequency at which the speaker cone resonates. This specification is given for woofers only and is useful in designing bass-reflex speaker enclosures.

Moving mass is the effective mass of all the moving parts of the loudspeaker. It is given for woofers and some midrange speakers. The moving mass and the speaker compliance (or stiffness) determine the free-air resonant frequency.

The *Q* of a speaker denotes its resonance magnification, which takes into account the damping of the speaker and the tendency of the speaker to reach its maximum sound output level when operating at the free-air resonance frequency. The Q and resonant frequency are increased when the speaker is placed in a sealed enclosure. Q is reported for woofers only and is in the range of 0.2 to 1.5.

The *V* or *compliance* of a speaker is given for woofers only. The compliance of a speaker is the volume of air (hence denote V) which has the same compliance as the speaker's suspension.

The *sensitivity* or *SPL* (sound pressure level) is the volume of sound produced by the speaker when it is fed one watt of electrical power within its frequency range. The SPL test is usually made with a microphone placed one meter from the speaker. An eight-inch woofer may have an SPL of 88 dB while a 15-inch woofer may have an SPL of 95 dB.

Power rating is the maximum watts RMS that a speaker can safely handle. If two numbers are given, the lower number is the maximum power that can be continuously fed to the speaker. The higher number is the maximum power that can be fed to the speaker for short periods of time without causing damage. A speaker with a high power rating has a large voice coil which allows more heat to be dissipated.

Magnet weight is the weight of the loudspeaker's permanent magnet. The magnet weight affects the damping and the efficiency of the loudspeaker. Woofers need large magnets because the speaker cone must move large distances to reproduce low frequency, high volume sounds. High power woofers can have magnets as heavy as fifty ounces.

The loudspeaker cone can be made of paper, cloth or polypropylene plastic. Cloth and/or paper cones are not as durable as polypropylene plastic cones.

The loudspeaker *suspension* determines the compliance of the speaker. Suspensions can be made from folded paper, rolled rubber or rolled polyfoam. Folded paper suspensions are stiff and are therefore suitable for use in ported reflex speaker enclosures. Rolled polyfoam or butyl rubber suspensions permit the speaker cone to move more freely, resulting in a high compliance speaker suitable for use in an acoustic suspension (sealed box) speaker enclosure.

ELECTROACOUSTICAL ANALOGY

An acoustical system may be represented and analyzed by its equivalent electroacoustical analogue. The electroacoustical analogues are easier to construct than models of the corresponding acoustical system; therefore, experimental results are more conveniently obtained from the electroacoustical analogue.

The acoustical and electrical systems are analogous if their differential equations of motion are mathematically the same; that is, the corresponding terms of the differential equations of motion are analogous to each another. There are two electrical analogies for acoustical systems: the *voltage-pressure* analogy and the *current-pressure* analogy, as given in *Table 6-1*.

Acoustical inertance is the ratio of acoustic pressure to the rate of change of the volume velocity:

$$M = P/(dV´/dt) \text{ kg/m}^4$$

Acoustical resistance is the ratio of acoustic pressure to the volume velocity:

$$R = P/(dV/dt) \text{ nt - sec/m}^5.$$

Acoustical compliance is the ratio of volume displacement to the acoustic pressure:

$$C = X/P \text{ m}^5/\text{nt}.$$

Acoustical System	Electrical System	
	Voltage-Pressure	Current-Pressure
P Pressure (nt/m^2)	E Voltage (volts)	I Current (ampere)
M Intertance (kg/m^4)	L Inductance (henry)	C Capaticance (farad)
X Volume displacement (m^3)	Q Charge (coulomb)	S(vdt) impulse (volt-sec)
X´ Volume velocity (m^3/sec)	I Current (ampere)	V Voltage (volt)
R Resistance (nt-sec/m^5)	R Resistance (ohm)	1/R Conductance (mho)
C Compliance (m^5/nt)	C Capacitance (farad)	L Inductance (henry)
Z Impedance (ohm)	Z Impedance (ohm)	1/Z Admittance (mho)

Table 6-1.
Electrical analogies for
acoustical systems.

Problems

6-1. What are the similarities and differences between a loud-speaker and a microphone?

6-2. Name two types of microphones. Give examples of each type.

6-3. Calculate a carbon microphone's sensitivity and response:

Where: $E = 12V$, $h = 7.5 \times 10^8$ ohm/m, $A = 0.000314$ m^2, $R = 120$ ohms, and $s = 10^6$ nt/m.

6-4. How can the resonance of a microphone be controlled?

6-5. What makes up the driver of a dynamic microphone?

6-6. Are the impedance and the DC resistance of a dynamic speaker the same? Why?

6-7. How do dynamic and electrodynamic speakers differ?

6-8. What type of speaker has a high impedance?

6-9. Why is an electrostatic speaker not suitable as a woofer?

6-10. Is a crystal speaker useful as a woofer? Why?

6-11. Calculate the acoustical power output of a speaker

Where: $B = 1$ weber/m^2, $L = 7.5$ m, $R = 2$ kg/sec, $E = 20V$, $Z_m = 13.3$ kg/sec and $Z_i = 11.7$ ohms.

6-12. Define a woofer and a tweeter.

6-13. Define free-air resonance.

6-14. Why do woofers need large magnets?

6-15. What determines a speaker's compliance?

6-16. Name two electrical analogies for acoustical systems.

Chapter 7

Noise, Music and Speech

There are three types of sound; namely, noise, music and speech. Noise is any non-periodic or non-repetitive vibration, as shown in *Figure 7-1*. Music is any periodic or repetitive vibration, as shown in *Figure 7-1*. Human speech is a natural sound source or sound transmitter.

NOISE

Noise is subjectively defined as any unpleasant or unwanted sound. Noise is objectively defined as the combined result of single-frequency sounds or pure tones, and it has a continuous frequency spectrum of irregular amplitude and irregular waveform, as shown in *Figure 7-1*.

Airborne noise is caused by variations in the air pressure. *Structural-borne* noise is caused by mechanical vibrations of elastic bodies. *Liquid-borne* noise is caused by variations in the liquid pressure. *Ultrasonic* noise is sound above the limits of human hearing, while *infrasonic* noise is sound below the limits of human hearing.

Noise pollution interferes with work, recreation and sleep. It causes strain, fatigue, loss of appetite, indigestion, irritation and headaches. High intensity noise has an accumulative adverse effect on the human ear. After prolonged periods of high intensity noise, temporary or even permanent deafness may result. Noise has a

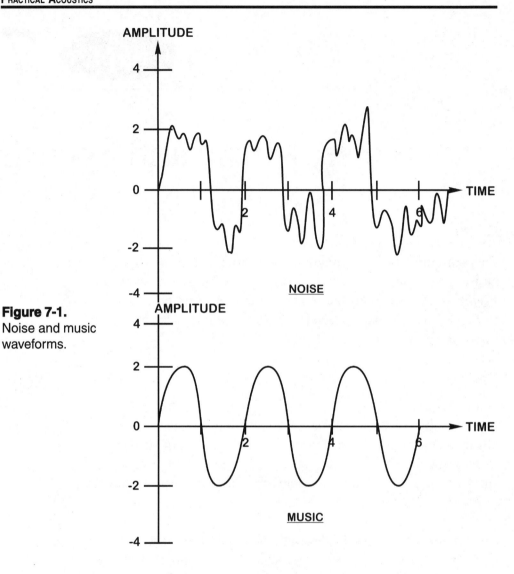

Figure 7-1.
Noise and music waveforms.

psychological effect on workers. Their output is lowered and their error rate increases because they are distracted from their work. Excessively loud, low pitched sounds cause muscles to tighten in the eardrum. This is stressful to the eardrum. Sustained loud noise can wreck the internal organs of a mouse and eventually kill it.

LOUDNESS

The loudness of a sound is the magnitude of the auditory sensation produced by the amplitude of the disturbance reaching the ear. Loudness is a subjective interpretation and therefore cannot be measured. Loudness is measured on a relative scale which is based on the logarithm of the ratio of two intensities, one of which is the reference intensity.

The *sone* is an acoustic unit used to measure the loudness of a sound. One sone is a pure tone of 1000 hertz at a sound intensity level of 40 decibels. One millisone corresponds to the threshold of hearing. A loudness of two sones is twice as loud as a loudness of one sone.

The *phon* is an acoustic unit used to measure the overall loudness level of a noise. One phon is a pure tone of 1000 hertz at a sound intensity level of one decibel. Other tones have a loudness level of N phons if the ear interprets them as sounding as loud as a pure tone of 1000 hertz at a sound intensity level of N decibels.

A tone with a loudness level of 20 phons does not sound half as loud as a tone with a loudness level of 40 phons. A tone of frequency 300 hertz at a loudness level of 20 phons sounds exactly as loud to the ear as any other 20-phon tone at any other frequency.

Loudness level of a sound is:

$$LL = 10\log \left(\frac{I}{10^{-12}}\right) \text{ phons}$$

where: I is the sound intensity in watt/m^2.

At low intensity sound levels, the human ear is most sensitive to frequencies between 1000 and 5000 hertz, as shown in the Fletcher-Munson curves of *Figure 4-1*. At very high intensity sound levels, the response of the human ear is more uniform, per *Figure 4-1*.

The *noy* is an acoustic unit used to rank the annoyance of noises to the human ear.

Perceived noise level is a subjective scale used to measure the unwantedness of a noise. Perceived noise level represents the intensity of noise and its frequency spectrum.

NOISE ANALYSIS

The overall sound pressure level of a noise is measured with a sound level meter and a sound analyzer. An audio frequency spectrometer and a level recorder are used to plot the pressure spectrum level of a noise over the band of audible frequencies.

An *octave* is the interval between two frequencies having a ratio of 2:1. A *one-third* octave band is a band of frequencies in which the ratio of the extreme frequencies is equal to the cube root of two; that is, 1.26. A *narrow* band is a band whose width is less than one-third octave but not less than one percent of the center frequency.

Intensity spectrum level at any frequency, F, of a noise is the intensity level of the noise contained within a band of frequencies one hertz wide, centered on the frequency F:

$$\text{ISL} = 10\log(\frac{I}{I_o dF}) = \text{IL} - 10\log(dF) \text{ dB}$$

where: I is sound intensity watt/m^2, I$_o$ is the reference intensity, IL is intensity level in dB, and dF is the bandwidth in hertz.

Pressure spectrum level is the sound pressure level contained within a band of frequencies one hertz wide:

$$\text{PSL} = \text{SPL} - 10\log(dF) \text{ dB}$$

where: SPL is sound pressure level in the band dF wide.

Pressure band level is:

PBL = PSL + 10log (dF) dB

White noise has a constant spectrum level over the band of audible frequencies. White noise may or may not be random and it may or may not be time dependent. The amplitude of a random noise occurs as a function of time according to the Gaussian distribution curve. A random noise does not have a uniform frequency spectrum. *Pink noise* has equal energy per octave from 20 to 20,000 hertz.

MEASURING ACOUSTICAL NOISE

Acoustical noise may be measured by using a condenser microphone to pick up the acoustical noise. The equipment setup is shown in *Figure 7-2*. The microphone converts the acoustical noise into an electrical signal. The tuned amplifier may be deleted if only the peak noise amplitude is being measured.

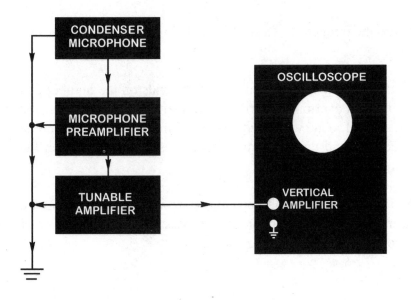

Figure 7-2.
Measuring acoustical noise.

MUSIC

Music is a subjective and complex mental sensation which occurs when somebody listens to a combination of sounds produced by vibrating strings, membranes and air columns. Musical tones have a simple harmonic structure with regular waveforms. They consist of fundamental and harmonic frequencies of integral-related frequencies. Musical acoustics involve physical and psychological laws as well as aspects of tone production. Musical sounds differ from each other in intensity, pitch, timbre or tone quality, and loudness.

Intensity is the amplitude of the sound wave or music source vibration transmitted per unit area. The intensity of a sound is directly proportional to the square of the amplitude of the sound wave. The intensity is inversely proportional to the square of the distance between the sound source and the sound receiver. The intensity of sound is directly proportional to the density of the transmitting medium. The *mel* is an acoustic unit which describes the pitch of a sound. A pure tone of 1000 hertz with a loudness level of 40 phons has a pitch of 1000 mels.

Pitch is the frequency of a sound. Pitch describes how low or how high a sound is.

The *timbre* or tone quality of sound depends on the number and amplitude of overtone or harmonic frequencies that occur with the fundamental sound wave frequency. The intensity of the harmonic frequencies and the harmonic frequencies themselves affect the timbre of a sound.

The loudness of a sound depends on how well the ears and the hearing mechanism function. It also depends on the intensity of the sound waves striking the eardrums.

MUSICAL NOTES AND SCALES

The current even-tempered musical scale is a compromise. Johann Sebastian Bach proposed the even-tempered scale which has been universally adopted, except in Oriental countries. The octave is divided into twelve equal intervals. The frequency of any note in this scale is determined by multiplying the frequency of the adjacent lower note by the twelfth root of two; that is, 1.06. The even-tempered scale is represented by eight white keys and five black keys in an octave on the piano. The new notes needed in addition to the eight fundamentals of the diatonic scale are represented by black keys, each of which serves as the sharp of the note below it and the flat of the note above it.

Doubling the frequency of a note raises its pitch one octave. A *major triad* is three frequencies with the ratio 4:5:6. Notes C, E and G are a major triad. A *major chord* is four frequencies with the ratio 4:5:6:8. The fourth note is one octave higher than the first note. Notes C, E, G and C´ are a major chord, where note C´ is one octave higher than note C. A *keynote* is the first note of any octave. Music is not always written in the key of C. Another note can be selected as the keynote of the scale. The musical scale is derived from three triads. A major diatonic scale is given in *Table 7-1*.

Frequency	Triad	Note	Syllable
24	1	C	do
27	-	D	re
30	1	E	mi
32	3	F	fa
36	1, 2	G	sol
40	3	A	la
45	2	B	ti
48	3	C´	do´
54	2	D´	re´

Table 7-1.
Major diatonic scale.

The *sixth overtone* does not correspond to any note in the scale because it generates an unpleasant sound. In order to avoid the sixth overtone, piano keys are adjusted to strike the strings at a place which prevents the strings from vibrating in seven segments.

Beats occur when two sound waves of slightly different frequencies alternately reinforce and cancel each other. Beats can cause discord or unpleasant sounds. Beats below 60 hertz are unpleasant sounds, especially at 30 hertz. Beats above 60 hertz are pleasant sounds.

MUSICAL INSTRUMENTS

The oldest reference to musical instruments occurs in the Old Testament. Around 4000 B.C., Joshua used high intensity sound waves to knock down the walls of Jericho.

There are three types of musical instruments, the *wind* instrument, the *string* instrument and the *percussion* instrument.

The wind instrument employs vibrating air columns to produce musical sounds. Varying the length of an air column varies the pitch of the sound. A shorter air column produces a higher frequency note. A longer air column produces a lower frequency note.

The string instrument uses vibrating strings to produce musical sounds. The frequency of a vibrating string is inversely proportional to its length. Shorter strings vibrate at higher frequencies. The heavier a string is per unit length, the more slowly it vibrates. The frequency of a vibrating string is inversely proportional to its length, diameter and to the square root of its density. The frequency of a vibrating string is directly proportional to the square root of its tension. Shorter, tighter and lighter strings produce high frequency notes. Longer, looser and heavier strings produce low frequency notes. Strings can vibrate as a whole or in parts. A stretched string plucked at its midpoint vibrates at its lowest or fundamental frequency. If a string is pressed at its midpoint and plucked halfway between one end and the midpoint, it vibrates at one octave above its fundamental frequency; that is, at double its fundamental frequency. Sympathetic or resonant vibrations intensify sound. Forced vibrations also intensify sound. Forcing the body of a violin to vibrate intensifies the sound of the strings.

Percussion instruments are struck to produce musical sounds. Percussion instruments produce many types of musical sounds.

SPEECH

Speech sounds are complex acoustic waves that provide listeners with many clues. Speech concerns the structure of language and it is characterized by the interpretive aspect, loudness, pitch, tempo and timbre. The intelligibility of speech is an indication of how well speech is recognized and understood and it depends on the acoustic power delivered during speech, speech characteristics, hearing acuity and ambient noises.

Sound articulation is the percentage of the total number of speech sounds correctly identified. *Syllable articulation* is the percentage of the total number of syllables correctly identified. Articulation increases rapidly with increasing speech level until it is 70 decibels.

Speech interference level, or SIL, is the average of readings in the three octave frequency bands of 600-1200 Hz, 1200-2400 Hz and 2400-4800 Hz. SIL is measured in decibels. A *voice speech spectrogram* is a time series of frequency versus amplitude plots.

The *masking* of a host sound is the shift in the threshold of hearing of the host sound due to the masking sound. It reduces the ability of a listener to hear the host sound. Degree of masking at any frequency is the difference in decibels between the background noises and the normal threshold of audibility.

Pure tones can be used as the masked sounds. A high pitch tone can mask a low pitch tone. A continuous bland background noise dulls the edges of an intermittent harsh sound.

HUMAN VOICE MECHANISM

Humans communicate their thoughts and ideas with speech. The human voice mechanism is a very low-efficiency sound producing system. The voice mechanism has a power generator, a vibrator, a resonator and an articulator. The anatomy of the speech mechanism is shown in *Figure 7-3*. The power generator includes the diaphragm, lungs, bronchi, trachea and associated muscles.

The *vibrator* is the larynx. The *larynx* is a cartilaginous box that rests below the base of the tongue and above the trachea. This positioning of the larynx places it in the direct air flow of everyday breathing. Every breath to and from the lungs must pass through the larynx. The larynx automatically moves up and down whenever you swallow, pressing against the epiglottis, and closing off the air passage. The *epiglottis* is a small piece of cartilage that acts like a lid over the larynx. It keeps food from entering the trachea or windpipe.

The *resonator* includes the nose, mouth and throat which resonate. The chest, head and palate serve as sounding boards. The mouth and adjacent cavities of the nose and throat constitute an adjustable amplifier for the sounds produced by the vocal cords. The amplifier operates on the principle of resonance and it can be tuned to a wide range of pitches by varying the positions of the tongue, lips and jaw. The resonator produces vowel sounds.

The *articulators* include the lips, tongue, teeth and palate. Articulators manipulate the air flow. They produce consonant sounds.

The vocal cords are located in the larynx. The *vocal cords* are two elastic membranes whose thickness, length and tension affect the pitch of the voice in response to the will of the person and in keeping with the maturity and sex of the individual. The vocal cords produce speech and music when air is forced up from the lungs through the glottis, causing the vocal cords to vibrate. These vibrations result in periodic pressure variations that give rise to speech waves.

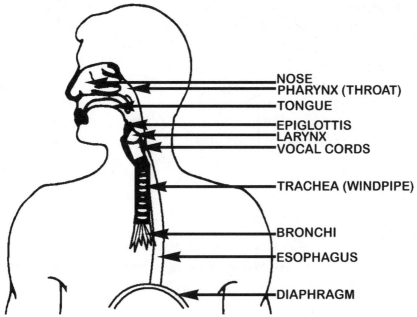

NOSE
PHARYNX (THROAT)
TONGUE
EPIGLOTTIS
LARYNX
VOCAL CORDS
TRACHEA (WINDPIPE)
BRONCHI
ESOPHAGUS
DIAPHRAGM

Figure 7-3.
Anatomy of the
speech mechanism.

The loudness of the human voice is determined by the stream of air forced from the lungs through the vocal cords. The pitch of the human voice is controlled by the elasticity and vibration of the vocal cords. The resonators control the timbre of the sound produced.

Vowels are musical sounds and are primarily produced by the vocal cords. Consonants are noise sounds and are primarily breath sounds controlled by the teeth, tongue and lips.

Laryngitis causes the vocal cords to become thicker. They vibrate more slowly than usual and the voice pitch is lower than usual. The voice is hoarse.

Short vocal cords produce high pitch sounds. Little children have short vocal cords and therefore have high pitch voices. In adolescence, the larynx enlarges and the vocal cords become longer, thicker and further apart. The voice becomes lower in pitch.

When you whisper you are speaking without using your vocal cords. You are using the gentle hiss of exhaled breath, which is shaped, as in normal speech, by the teeth, tongue, lips and palate.

Your voice reaches your inner ear by two routes. It enters your eardrum when it travels by air. It enters your jaw when it travels through your jaw—the medium is bone. Your voice reaches other people only through the medium of air; therefore, your voice sounds quite different to you than it does to other people.

ELEMENTS OF SPEECH

The resonated and articulated vocal cord vibrations is human speech. There are five additional elements to good speech; namely, breathing, pitch, volume, velocity and timbre.

Breathing consists of inhalation and exhalation. Exhalation is important to speech because speech mastery is accomplished during exhalation. A controlled breath is required for maintaining accurate vowel and consonant phonation throughout the duration of the spoken noise, music or speech. Inhalations can be obtained during the natural pauses of speech.

The *pitch* of a voice is determined by the shape and size of the larynx. Inflection in speech is another aspect of the pitch of a voice. *Inflection* in speech can be the difference between a monotonic dialog and an enthusiastic oratory. Pitch can be used to indicate emotion and to signal correct pronunciation.

Volume is how loud you are speaking and it can indicate your emotions. An angry person talks louder than a relaxed person. Physical disabilities such as deafness can affect the volume of speech.

Good speech is composed of measured phonation bracketed by appropriate pauses. Short pauses indicate a rapid speech pattern and long pauses indicate a slow speech pattern.

The *timbre* of a voice is governed by its overtones. Harmonic tones serve as the unique signature for an individual's voice. Twang, nasal, lisp and drawl are different speech qualities. *Twang* is nasal speech. Nasal speech is speaking with the voice passing through the nose. *Lisp* is a speech defect where sibilants "S" and "Z" are articulated like "TH" in thank and "TH" in this. A *sibilant* is a consonant produced by the frictional passage of breath through a narrow opening in the front part of the mouth. *Drawl* is speaking slowly. Like other elements of good speech, voice quality can be altered for a more desirable timbre.

Phonetics is the study of speech sounds. *Phonemes* are the smallest units of speech. Phonemes constitute the differences found in the diverse languages of the world. Continuants, stops and glides are three types of speech movements in phonetics.

Continuants are the prolonged holding to an initial phoneme. Vowels, fricative consonants and nasals are continuants. Vowels are open sounds which are amplified by the vocal mechanism's resonators. A, E, I, O and U are vowels. *Fricative consonants* are articulated through a frictional passage. F, V, TH, S, Z, SH and H are fricative consonants. *Nasals* are vowels or consonants exhaled through the nose. M, N and NG are nasal consonants.

Stops are the blocking and unblocking of the air flow through the larynx. *Plosive consonants* are formed by completely stopping the articulation of a consonant. B, D, P and T are plosive sounds.

*Glide*s are the continuous movement of the vocal mechanism from one phoneme to another phoneme. Glides are composed of *dipthongs* which are formed by gliding from the articulatory position of one vowel to the articulatory position of another vowel. AU and OI are dipthongs.

Problems

7-1. What is the difference between noise and music?

7-2. What is the difference between ultrasonic noise and infra-sonic noise?

7-3. What type of sounds cause muscles to tighten human ear drums?

7-4. Can loudness be measured? Why?

7-5. Is the human ear equally sensitive to all frequencies at low sound intensities? If not, what frequencies are they more sensitive to at low sound intensities?

7-6. The sound intensity of each one-hertz band of noise is:

$$\frac{10^{-5}}{F} \text{ watts/m}^2$$

where: F is the center frequency of the band.

What is the intensity spectrum level of the noise at 2000 Hz? What is the intensity level of noise between 1500 Hz and 2500 Hz?

7-7. What is the difference between white noise and pink noise?

7-8. How do musical sounds differ from each other?

7-9. What determines the timbre of a sound?

7-10. Define a major triad.

7-11. Define a major chord.

7-12. Define a keynote.

7-13. Define a beat.

7-14. Name three types of musical instruments.

7-15. Where are the human vocal cords located?

7-16. Describe vowel and consonant sounds.

7-17. Why do people have hoarse voices when they have laryngitis?

7-18. Does your voice sound the same to you as to other people? Why?

7-19.   Name five elements of good speech.

7-20. Name some speech qualities.

7-21. Define phonetics.

7-22. Define a phoneme.

Chapter 8

The Human Ear

The human ear is a sound receiver because it translates acoustic pressure fluctuations into pulses in the auditory nerve. These pulses are sent to the brain, which interprets them and converts them into sensations; that is, the perception of sound. *Sound* is a physiological sensation which involves the ears and the brain.

The frequency response of the human ear is a subjective quantity. It cannot be measured directly. The response of the human ear varies with frequency and sound intensity. The human ear is more sensitive to changes in frequency than to changes in sound intensity. It is more sensitive to low intensity sounds than it is to high intensity sounds. The human ear creates sounds of various frequencies because it has a nonlinear response to sound waves. The frequency range of the human ear is 20 to 20,000 hertz. As we age, the frequency range of the ear decreases.

Hearing loss, HL, is the decibel difference between a subject's threshold of audibility and that for a person with normal hearing at a given frequency. It is a shift in sensation level:

$$HL = 10\log\left(\frac{I}{I_o}\right) \text{ dB}$$

where: I is threshold sound intensity for the subject's ear, and I_o is the threshold sound intensity for the normal ear.

Sensation level, SL, of a tone is the amount it exceeds the threshold of hearing, in decibels:

$$SL = 10\log \left(\frac{I}{I1}\right) dB$$

>where: I is the intensity of the tone and where I1 is the intensity of the threshold of hearing at that frequency or tone.

The hearing mechanism is resilient to changes in intensity. The human ear can be overloaded. *Deafness* is the amount of hearing loss in decibels. *Conduction deafness* is hearing impairment due to an abnormality or obstruction of the middle ear. *Nerve deafness* is hearing loss due to a nerve defect or damage.

Tone deafness is the inability to discriminate small differences in pitch. Tone deafness is not inherited. It is an educational or environmental handicap. Tone discrimination is taught at an early age.

Hearing tests employ an audiometer, an attenuator, an interrupter switch and an earphone. The hearing test is used to determine the threshold of hearing, hearing defects and hearing deterioration.

The subject wears headphones and listens to several single-frequency tones. Each tone is introduced at an inaudible level and increased in loudness until the subject indicates that he can hear the tone. This tests for conduction deafness.

In a different test, the subject wears earphones and tones are transmitted to the bones of the skull. The subject indicates if he can hear the tones. This tests for nerve deafness.

The two human ears can identify and locate the direction of a sound source with great accuracy. This ability is called *binaural audition* or *auditory localization* and it is due to the difference in sound intensity at the two ears, due to diffraction at the two ears and due to the phase difference in sound arriving at different times at the two ears.

Frequencies less than 700 hertz are not blocked by the human head because the distance between the two ears is half a wavelength of the sound wave or less. These frequencies are localized by the phase difference between the human ears.

In the range of 700 to 5000 hertz the directional behavior of the energy field around the listener is important. The head is an obstacle because the wavelength of sound waves in this frequency range is less than the diameter of the human head. Head movement is used by the brain to estimate the probable subjective mid-frequency and high-frequency sound direction.

Sounds of frequencies above 5000 hertz are located by the *pinnae* or flaps of the ears, which modify sounds from different directions. The pinnae's acoustic obstruction gives good spatial localization. The pinnae localization mechanism appears to rely on the fact that sound from each direction arrives inside the listener's ear with a distinctive coloration.

ANATOMY OF THE HUMAN EAR

The human ear is a miracle of miniaturization. It is the size of a hazelnut. It contains the receptors of the eighth cranial nerve. The ear deals with hearing and balance. It has three parts; namely, the external ear, the middle ear and the inner ear. The anatomy of the human ear is shown in *Figure 8-1.*

The *external* ear consists of the pinna and the ear canal. The pinna is a cartilage framework covered with skin and it projects from each side of the head. The pinnae collect and direct sound waves into the ear. The external acoustic meatus or ear canal is about one inch long. It has a double curve and it is lined with skin that has fine hairs, oil or sebaceous glands and wax or ceruminous glands.

The *middle ear,* or tympanic cavity, consists of the eardrum, the auditory tube, the hammer, the anvil, the stirrup, and the oval window. The middle ear is located in a hollowed-out area of the temporal bone. The auditory tube connects the ear to the throat. The

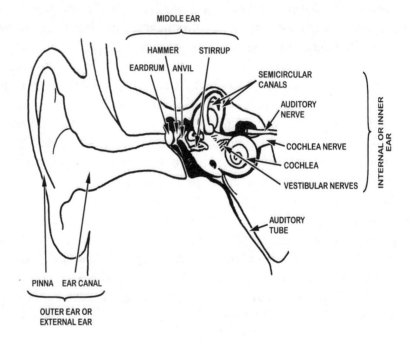

Figure 8-1.
Anatomy of
the human ear

tube helps to equalize the air pressure in the cavity with that outside the body by opening when the person swallows. The entrance to the middle ear is guarded by the eardrum, which is a fibrous membrane. The roof of the cavity is formed by a thin plate of bone which separates it from the middle cranial fossa. Thin bone also forms the floor of the cavity, which separates it from an opening in the floor of the skull. The oval window is the opening at the other end of the cavity.

There are three tiny bones, or ossicles, suspended by ligaments from the walls of the cavities. The three bones, the *hammer*, *anvil* and *stirrup*, form a chain across the middle ear. The handle of the hammer is connected to the eardrum. The foot-plate of the stirrup is connected to the oval window. The anvil lies between the hammer and the stirrup. The *tensor tympani* is a small muscle which extends from the auditory tube to the handle of the hammer. When this muscle contracts, the membrane of the eardrum is tightened. The *stapedius* is another tiny muscle which is attached to the stirrup. When the stapedius contracts, the foot-plate is tilted.

The three bones are connected by true joints and the bones form a lever system that transmits sound waves through the middle ear. Sound wave energy rocks the foot-plate of the stirrup in the oval window. Sound vibrations are transmitted across the middle ear, through the oval window and into the internal ear.

The *inner ear* consists of the semicircular canals, the vestibule and the cochlea. The bony labyrinth is lined by a membranous labyrinth. The *vestibule* is the main chamber of the labyrinth. The vestibule has three semicircular canals opening into its posterior wall. The *semicircular canals* are at right angles to each other and they are instrumental in maintaining the balance or equilibrium of a person in motion. The round and oval windows are at either end of the vestibule.

The *cochlea* is a snail-shaped canal which winds two and one-half times around a pillar of bone. The cochlea opens into the wall of the vestibule. The cochlea is composed of three connected compartments. The compartments are the *upper scala vestibuli*, which is bounded laterally by the oval window, the middle *cochlea duct* and the lower *scala tympani*, which is bounded laterally by the round window.

The spiral *organ of Corti* is the organ of hearing and it is contained within the cochlea. The Corti has a basilar membrane which forms the floor of the cochlear duct and separates the cochlear duct from the scala tympani. Resting on the basilar membrane are columnar cells with cilia or hair cells that are in contact with fibers of the cochlear nerve.

The basilar membrane has at least 20,000 separate fibers which are attached to the auditory nerve. The *auditory nerve* is the pathway to the brain of sound sensations. Each fiber in the basilar membrane is associated with a particular frequency of sound wave.

The membranous labyrinth and the bony labyrinth are separated by the perilymph fluid. The endolymph fluid fills the membranous channels.

PHYSIOLOGY OF HEARING

The ear changes sound waves into nerve impulses. Sound waves are directed into the ear canal by the pinnae. The sound waves strike the eardrum, causing it to vibrate. The mechanical energy of these vibrations is transmitted across the middle ear by the hammer, anvil and stirrup. The stirrup rocks against the oval window which causes the perilymph in the membranous cochlea to be set in motion. The wave begins in the scala vestibuli, proceeds through the cochlear duct, continues into the scala tympani, and finally the wave pushes against the round window.

The cochlea converts the sound waves into nerve impulses. The basilar membrane, when set into motion by the fluid wave, initiates the action of the hair cells resting on it. The fibers of the cochlear branch of the eighth nerve form a network around the hair cells. Movements of the basilar membrane set up nerve impulses in the hair cells. The basilar membrane detects the pitch and amplitude of the sound and it transmits this information, via the hair cells, to the cochlear nerve fibers. The auditory pathway by which the nerve impulses reach the cerebral cortex is complex and it involves at least four neurons. Sound perception and interpretation occur in the brain.

Problems

8-1.   Find the sensation level of a tone of intensity 10^{-6} watt/m² and frequency 50 hertz. The threshold of hearing at 50 hertz is 10^{-7} watt/m².

8-2.   Define conduction deafness. Define nerve deafness.

8-3.   Define tone deafness.

8-4.   Define binaural audition.

8-5.   How are sounds below 700 hertz located?

8-6.   How are sounds in the frequency range of 700 to 5000 hertz located?

8-7.   How are sounds above 5000 hertz located?

8-8.   How large is the human ear?

8-9.   What is the function of the human ear?

8-10.   What is the function of the cochlea?

8-11.   What is the function of the basilar membrane?

8-12.   Where do sound perception and interpretation occur?

Chapter 9

Architectural Acoustics

Architectural acoustics deals with reverberation control, noise insulation and reduction, sound absorption and sound distribution. Good architectural acoustics results in the intelligibility of speech, in the reduction of external noises and in the richness of music.

Reverberation is the persistence of sound in a room because of continuous reflections of sounds at the walls once the sound source is turned off. Reverberation depends on the size and shape of the room and on the frequency of the sound.

Wallace Sabine pioneered the study of reverberation. In 1900, he demonstrated that absorbing materials can control excessive reverberation. The newly built lecture hall at Harvard University had excessive reverberation. The spoken word would echo for about five and one-half seconds. Professor Sabine borrowed about 1500 seat cushions and he spread them around the new lecture hall. The results were amazing.

The duration of audibility for a sound from an organ pipe dropped from about five and one-half seconds to about one second. The lecture hall walls were promptly lined with felt for a permanent solution. Reverberation control by sound absorption was established by this simple experiment; that is, placing seat cushions around the new lecture hall at Harvard University.

The best speech acoustics are obtained when the reverberation time of a room is about one second. The best music acoustics are obtained when the reverberation time of a room is at least two seconds. For music, some reverberation is desirable so that musical sounds can blend and reverberate.

There are two types of sound incident on a listener in a concert hall; direct sound and reverberant sound. *Direct* sound arrives at the listener's ears directly from the sound source. *Reverberant* sound arrives after one or more reflections from either the surfaces of the concert hall or from objects inside the concert hall.

Reverberation time at a specific frequency is the time in seconds for the sound pressure to decrease by 60 decibels of its original value after the sound source is turned off:

$$T = 0.161 \ V/a$$

> when: V is the room volume in cubic meters, and a is the total room absorption in sabins.

Similarly:

$$T = 0.049 \ V/a$$

> when: V is the room volume in cubic feet, and a is the total room absorption in sabins.

When the reverberation time of a room is too short, the sound may not be sufficiently loud in all areas of the room. When the reverberation time of a room is too long, echoes are present. Good speech intelligibility requires a reverberation time of about one second. If the reverberation time of a room is less than one second, the sound intensity in the room is decreased and speech intelligibility is therefore decreased. Reverberation time is an important measure of good architectural acoustics.

A *reverberation chamber* is a room constructed with paddle-like turning vanes which cause uniform sound diffusion. The room surfaces have almost no sound absorption. The walls reflect sound waves, which suffer very little loss at each reflection. The sound energy distribution at any point in the room (not too close to the walls or the sound source) is uniform. The sound appears to be coming equally from all directions.

For any sound source, the level of the reverberant field changes with the volume of the room and with the acoustic properties of the materials in the room. The level of the reverberant field decreases with increasing room volume. It increases with increasing absorption of sound by the walls and objects in the room. The distance from the sound source at which the sound pressures of the direct field and reverberant field are equal becomes larger.

Tiled showers often have reverberation times that are longer than those of a concert hall. We can tell the difference between these two environments because the times between the reflections of sound are important parameters for judging the size of a room. The average time between reflections in a concert hall is an order of magnitude greater than that of an average living room.

A reverberation chamber is used to measure the total sound power output of equipment, to establish the noise reduction coefficient, to calibrate microphones and to test the sound control efficiency of materials and structures.

The *growth* of sound intensity in a reverberation chamber is:

$$I(t) = \frac{W(1 - e^{-(ac/4V)t})}{a} \text{ watt/m}^2$$

Similarly, the *decay* of sound intensity in a reverberation chamber is:

$$I(t) = 0.25Ece^{-(ac/4V)t} \text{ watt/m}^2$$

where: W is the sound power output in watts, a is the total sound absorption in sabins, c is the speed of sound in m/sec, V is the volume of the room in m^3, and E is the sound energy density in joules/m^3 when the sound source is turned off.

NOISE INSULATION AND REDUCTION

Noise insulation or soundproofing is often required to reduce noise. Noise reduction is accomplished either by absorbing the noise or by reducing the transmission of noise.

Airborne noise enters buildings through holes, cracks, poorly fitting windows and doors, air intakes and air exhausts. Airborne noise can also cause walls to vibrate. Airborne noise can be reduced by breaking its transmission path, by using absorptive materials and by surrounding the noise source with sound barriers or silencers. Fiberwool insulation fitted between the wall studs for thermal insulation also effectively absorbs airborne sounds. Background noise can be reduced by using the acoustical treatments described to reduce airborne noise.

Transmission loss is the difference in decibels between the sound energy striking the surface separating two spaces and the sound energy transmitted. Transmission loss is airborne noise reduction and it cannot be measured directly. It is computed from sound pressure levels on both sides of the surfaces:

$$TL = 10 \log(\frac{S}{S_T}) = SPL1 - SPL2 \text{ dB}$$

where: S is the total area of the surface, and S_T is the total sound transmission coefficient.

Structural-borne noise is the result of vibrating elastic bodies. It can travel through walls, floors, columns, beams, pipes, ducts and other solid structures. Structural-borne noise carries more energy than airborne noise; therefore, structural-borne noise should be suppressed at its source. Its transmission paths should be interrupted by resilient mounting insertions such as rubber bushings and by sound plenums or traps. Walls should have discontinuities filled with air or absorptive materials.

Machine noise usually indicates poor balance, excessive clearance (usually caused by wear), turbulent flow or other improperly functioning component of the machine. Many machine noises can be reduced by repairing the machine. When necessary, acoustical filters such as mufflers, plenum chambers, resonators, hydraulic filters and sound traps can be used to reduce machine noise. Machine noise sources should be isolated and vibration-mounted to reduce noise and vibration transmission.

Impact noise such as footsteps can be reduced by using carpets to cushion the impact areas of floors. The floors can be isolated from supporting structures by resilient mountings such as rubber strips.

Space average sound pressure level is:

$$L = 10 \log ([P_1^2 + P_2^2 + ... + P_n^2]/nP_o^2) \text{ dB}$$

where: P_n are sound pressures in nt/m², and $P_o = 0.00002$ nt/m² is the reference sound pressure.

SOUND ABSORPTION

When sound energy is absorbed, part of its energy is converted into heat (by the frictional and viscous resistance of the pores and fibers of acoustical materials) and part of its energy is converted into kinetic energy (by the vibration of the acoustical materials).

Draperies, carpets, suspended space absorbers and moveable absorptive panels can be used to absorb unwanted sounds in rooms and buildings. Low frequency sounds are absorbed by thin panels with air trapped behind them. Helmholtz resonators and resonator-panel absorbers are useful for absorbing sounds that occur at their resonant frequencies. Mufflers impede the transmission of sound even though they permit the free flow of air.

The *sound absorption coefficient* of a material is the decimal fraction of perfect absorption that it has. It is the efficiency of a material in absorbing sound energy at a given frequency. It varies with the angle of incidence and with the thickness of the material. An open space has a sound absorption coefficient of unity. The sound absorption coefficient is obtained by averaging the ratio of absorbed to incident energy over all possible angles of incidence. The average sound absorption coefficient of a room is the average of the sound absorption coefficient of all the absorbing areas of the room.

Sound absorption is the total area in square feet of perfectly absorbing material. It is measured in *sabins*. One metric sabin is one square meter of perfect sound absorptive material.

The *noise reduction factor* is:

$$RF = TL + 10 \log(\frac{a}{S}) \ dB$$

where: TL is the transmission loss in decibels, a is the total sound absorption in sabins, and S is the area of the partition in square feet.

The *difference* in noise level is:

$$db_{before} - db_{after} = 10 \log \left(\frac{a_{after}}{a_{before}} \right) dB$$

SOUND ABSORBING ACOUSTICAL

Sound intensity in a room is inversely proportional to the amount of sound absorption present. If the room is very large and the total sound absorption is very small, then sound absorption in air is very important.

An anechoic chamber has highly absorptive wedges mounted to the walls of the room. All incident sound energy is absorbed. An *anechoic chamber* simulates a free field or unbounded space. Complete soundproofing of the anechoic chamber is achieved by building a floating floor inside the anechoic chamber. An anechoic chamber is useful for measuring the acoustic characteristics of equipment, for calibrating microphones and for measuring the sound radiation patterns of loudspeakers. The decay of sound in an anechoic chamber is:

$$I(t) = Ie^{(Sc/4V)\ln(1-a')t} \text{ watt/m}^2$$

where: I is the sound intensity when the sound source is shut off, S is the total wall area, c is the speed of sound, V is the volume of the anechoic chamber, and a´ is the average sound absorption coefficient of the anechoic chamber.

Sound Distribution

Sound distribution describes how the sound pressure level varies with the position in a room. Smooth growth and decay of sound is required for good architectural acoustics. Buildings and rooms are designed to have sound evenly diffused over the entire area by using acoustical treatments such as the scattering effect of objects within the room, wall surface irregularities, use of absorptive material, reflecting surfaces and diffusers.

Model analysis with light rays, ultrasonic waves or audio frequency sound waves is used to study sound distribution. Graphical construction of the first reflections of the sound waves at various cross sections of the room can also be used, as shown in *Figure 9-1*.

Room Acoustics

An acoustically well-designed room has an optimum reverberation time, which results in good intelligibility of sounds and sounds of sufficient intensity. The room has a minimum of extraneous and other unwanted sounds because it is well soundproofed. The room also has good sound distribution. An acoustically well-designed room should encourage oblique sound waves because they decay very rapidly. The room should discourage axial sound waves because they decay very slowly.

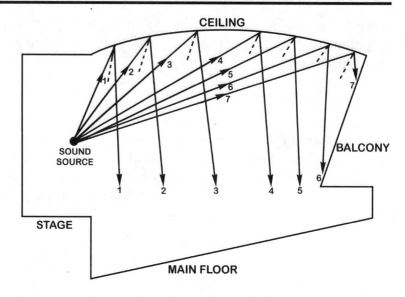

Figure 9-1.
Reflection of
sound waves.

<u>NOTES</u>
1. DASHED LINES ARE PERPENDICULAR TO THAT
POINT IN THE CEILING
2. ANGLE OF INCIDENT EQUALS ANGLE OF REFLECTION

Acoustic materials absorb sound energy. Curtains, drapes, carpets, clothing and upholstery are good sound absorbers. Hard surfaces such as glass and wood are good sound reflectors. Sound absorbers reduce reverberation in an acoustically well-designed room. Any room can be soundproofed by installing fiberwool thermal insulation between the wall studs and between the ceiling and/or floor joists.

An *echo* is heard if the reflected sound reaches the listener at least one-tenth second after the original sound reaches the listener. Sound travels 110 feet in one-tenth second; therefore, the reflecting surface must be at least 55 feet from the sound source to produce an echo.

Echoes, or reverberations, result when sounds reach a listener by two different paths that have vastly different lengths. Echoes are unpleasant in a concert hall. *Reverberation* is the multiple back-and-forth reflection of sound waves within a room. Some reverberation in a room is desirable. Excessive reverberation results in harsh sounds. Too little reverberation results in dull lifeless sounds. Reverberation time depends on the size of the room and whether the sound source is speech or music. For speech, the reverberation time of a room should be about one second. For music, the reverberation time of a room should be at least two seconds.

Room *flutter* occurs between two parallel walls that are smooth and highly reflective. Sound is reflected back and forth to produce multiple echoes.

Sound *focusing* is the concentration of sound at a point in a room due to the reflection of sound from curved surfaces. Sound distribution is unequal.

A properly shaped *reflector* can reflect sound waves such that they spread and come together somewhere else. A dome-shaped roof can be a sound reflector.

A *dead spot* is a region of sound deficiency due to the destructive interference of two or more sound waves.

An *acoustic shadow* is formed by an obstacle whose size is comparable with the wavelength of a sound wave. The sound wave is diffracted; that is, it is bent around the obstacle.

Percentage articulation is used as an intelligibility rating of rooms. Percentage articulation is determined from the shape of the room, from the noise in the room, from the reverberation time of the room, and from the sound intensity in the room.

The *room constant* is an alternate way to indicate and compare the acoustics of a room:

$$R = \frac{S}{1 - a'} \text{ ft}^2$$

where: S is the total wall area of the room in square feet
and a′ is the average sound absorption coefficient.

Problems

9-1.   Define reverberation.

9-2.   What type of materials are used to control reverberation?

9-3.   What is the ideal reverberation time of a room for speech? What is the ideal reverberation time of a room for music?

9-4.   What is the difference between direct sound and reverberant sound?

9-5.   Define reverberation time.

9-6.   A concert hall is 250 feet long, 100 feet wide and 20 feet high. The materials in the room have a total sound absorption of 10,000 sabins. What is its reverberation time?

9-7.   What happens to sound if the reverberation time of a room is too short? What happens to sound if the reverberation time of a room is too long?

9-8.   Define a reverberation chamber.

9-9.   A room has a volume of 90 cubic meters and a total sound absorption of 10 metric sabins. A sound source with 10 microwatts of sound power is turned on. If the speed of sound is 343 m/sec., what is the sound intensity at the end of 0.25 seconds?

9-10.   How does airborne noise enter a building?

9-11.   Define transmission loss.

9-12. What causes structural-borne noise?

9-13. Why should structural-borne noise be suppressed at its source?

9-14. What happens to sound energy when it is absorbed?

9-15. Define sound absorption coefficient.

9-16. Calculate the noise reduction factor if the area of a partition is 100 square feet. The transmission loss is 20 decibels and the total sound absorption is 10 sabins.

9-17. Define an anechoic chamber.

9-18. What happens to sounds in acoustically well-designed rooms?

9-19. What type of sound waves decay slowly? What type of sound waves decay quickly?

9-20. How far must a sound reflecting surface be from a sound source to produce an echo?

9-21. Define room flutter.

9-22. Define sound focusing.

9-23. Define dead spot.

Chapter 10

Underwater Acoustics

Underwater acoustics deals with the transmission of sound waves through water. There are transmission losses when sound waves travel through water. Sound waves are absorbed, reflected and refracted in water.

Water transmits sound waves much better than it transmits optical, radio and magnetic waves. Water temperature, pressure gradients, marine organisms, air bubbles and salt content affect the transmission of sound waves under water.

Sound transmission losses in sea water are due to divergence, absorption and irreversible attenuation. *Divergence* is an outgoing spherical acoustic wave whose decrease in intensity is inversely proportional to the square of its distance from the sound source. *Absorption* is the dissipation of acoustical energy into the medium or boundary because of viscous losses, heat conduction losses and molecular action. *Irreversible attenuation* is loss caused by refraction, scattering, diffraction and interference. Irreversible attenuation is also known as *transmission anomaly* and it is measured in decibels. The total transmission loss is:

$$H = 20 \log(r) + ar + A \text{ dB}$$

where: r is the distance in meters between the sound source and sound receiver, a is the absorption coefficient in dB/m, and A is the transmission anomaly in dB.

REFRACTION

Refraction is the bending of sound waves due to velocity changes that accompany pressure and temperature changes. Temperature varies linearly with the depth of the water; therefore, sound waves are refracted downward in an arc. Sound waves cannot reach the surface of the sea because sound waves are bent downward in water. A *shadow zone* is formed, as shown in *Figure 10-1*.

The temperature is constant at great depths below the sea. Sound velocity increases linearly with depth because pressure increases linearly with depth; therefore sound waves are refracted upward and follow an arc of a circle. At great depths *sound channels* are formed. Sound waves are refracted upward and downward along a narrow channel, as shown in *Figure 10-2*. Temperature and pressure gradients cause sound channels to be formed. Sound waves within the sound channels spread out in a circle rather than in a sphere. Sound waves therefore propagate to much greater distances.

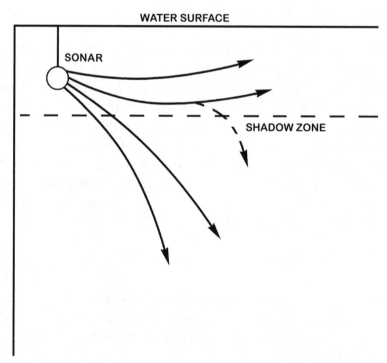

Figure 10-1.
Underwater shadow zone.

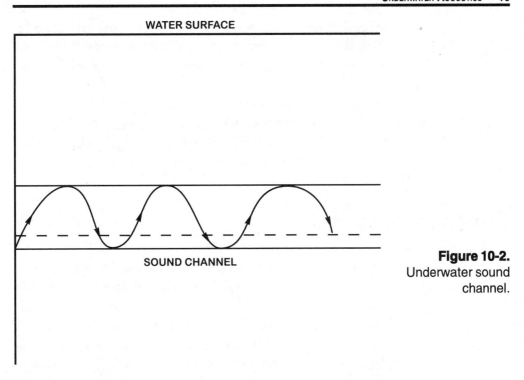

WATER SURFACE

SOUND CHANNEL

Figure 10-2.
Underwater sound
channel.

REVERBERATION AND AMBIENT NOISE

Reverberation, or *background scattering,* is transmitted acoustic energy that returns to the sound source without intercepting an object or target. Reverberation is directly related to the acoustic energy projected into the water by the sound source. Reverberation is usually an unwanted signal which tends to interfere with the returned echo.

Volume reverberation is caused by the scattering of sound in the bounded and non-homogeneous volume of the sea. *Surface reverberation* is due to reflections at the surface of the sea. *Bottom reverberation* is due to reflections at the bottom of the sea.

Echo-sounding is based on the reflection of sound and the production of an echo. Echo-sounding is used to locate submerged objects by emitting a sound wave and by receiving the returned echo. *Passive listening* is used to detect sounds from an unknown direction by collecting sound waves while maintaining complete silence. Passive listening has a longer detection range than echo-sounding and a passive listening system cannot reveal its location.

Ambient or background noise in the sea is unpredictable. Ambient noise in the sea is a function of wind, rain and the state of agitation of the sea. The desired signal is masked by biological noises, man-made noises, noises generated by ships and noises generated by echo-sounding and passive listening systems.

UNDERWATER TRANSDUCERS

A *hydrophone* is a transducer that converts sound waves in water into electrical energy. Ships produce sounds underwater. A hydrophone or underwater receiver picks up the sound from one direction only. The sound frequency transmitted should be *ultrasonic* (frequencies above 20,000 hertz) to prevent confusion with other sounds. Hydrophones are used to detect enemy ships in times of war.

Hydrophones should have good stability, high sensitivity and a linear frequency response. Their specifications must be independent of temperature. They must be able to withstand high hydrostatic pressures. Their faces must be large to meet high power and small displacement requirements.

Hydrophone *sensitivity* is the voltage generated at its terminals by sound pressure, in volts per microbar. It is a function of the angle measured from the acoustic axis (or the axis of maximum sensitivity) of the hydrophone and of the frequency of the signal generated.

Hydrophone *directivity* is an indication of the fraction of the total signal the hydrophone converts into electrical energy according to its sensitivity pattern. A hydrophone that is equally sensitive in all directions has a directivity factor of one and a directivity index of zero.

An underwater *sound projector* is an electroacoustic transducer that converts electrical energy into acoustical energy in the water by either piezoelectric or magnetostrictive effects. A piezoelectric crystal changes its length when a voltage is applied between two faces of the crystal. A magnetostrictive rod changes its length when it is exposed to a varying magnetic field.

CAVITATION

When the pressure of water is less than the vapor pressure of water, bubbles filled with water vapor are formed. The bubbles collapse when they move into a region of higher pressure. Their collapse or local boiling produces noise with vibration which impedes the transmission of sound. This phenomenon is known as cavitation. The cavitation number is:

$$CN = \frac{2(P_o - P_v)}{pV^2}$$

where: P_o is the ambient pressure, P_v is the vapor pressure of water, p is the density of water in kg/m^3, and V is the speed of the vehicle.

SONAR

Sonar systems are used to detect sounds in water. Sonar is an abbreviation for SOund Navigation And Ranging. It scans the water until its ultrasonic sound beam hits a target and produces an echo which is reflected back to the sonar system. A cathode-ray tube can be used to give a contour picture of the object detected. Sonar systems can be used to locate oil deposits. Salt domes above the oil deposits bend the sound waves which causes them to return to the surface of the water. The distance from the sonar system to the salt domes can then be calculated. There are two types of sonar systems, *active* sonar and *passive* sonar.

ACTIVE AND PASSIVE SONAR

Active sonar is a target-seeking system. It transmits underwater ultrasonic sounds that strike targets and return in the form of echoes which indicate the range and bearing of the target. Active sonar systems can be used to establish contact with targets, to determine the range and bearing of targets, and to determine the rate at which the range and bearing are changing. Active sonar is usually installed in surface ships. There are two types of active sonar, searchlight sonar and scanning sonar.

The *searchlight* sonar transmits at only one bearing at a time. The transducer is held at that bearing and it listens for returning echoes. The bearing is changed in steps to search the area around the ship. Searchlight sonar has a long range because the available power is focused into a concentrated beam.

Scanning sonar provides indications of all underwater objects around the ship. Its sound pulse spreads out in all directions simultaneously. The scanning sonar system has a limited range.

Some scanning sonar systems can transmit a directional beam throughout 360 degrees of rotation. This type of system increases the range dramatically.

Thermal layers in the ocean reflect sonar sound waves, allowing a submarine below the thermal layer to remain undetected. The *variable depth sonar* overcomes this problem. The transducer is placed in a towed vehicle. The towed vehicle is placed below the thermal layer and it is towed behind the ship. The transducer is connected by an electric cable to the shipboard sonar equipment. The effect of the thermal layer is minimized.

Passive or *listening* sonar systems do not transmit sound. It listens for sounds produced by the target to determine its range and bearing. Passive sonar systems are usually installed in submarines. Passive sonar systems use an array of hydrophones connected together and arranged in a circle or around the bow of a submarine.

Problems

10-1.   What can affect the transmission of sound waves underwater?

10-2.   What causes sound transmission losses in salt water?

10-3.   Why are sound waves refracted downward in an arc underwater?

10-4.   Why are sound waves refracted upward in an arc underwater?

10-5.   What causes ambient noises in the sea?

10-6.   What is a hydrophone?

10-7.   Define hydrophone directivity.

10-8.   What is an underwater sound projector?

10-9.   The ambient pressure is 400,000 nt/m^2. The vapor pressure of water at 20 degrees Celsius is 2400 nt/m^2. The density of water at 20 degrees Celsius is 1061 kg/m^3. Calculate the cavitation number if a boat travels at 15 m/sec in the water.

10-10. What are some applications of sonar?

10-11. What is the difference between active and passive sonar?

Chapter 11

Ultrasonics

Ultrasonics is the physics of sound waves having a frequency above the limits of human hearing; that is, above 20,000 hertz. The frequency range of ultrasonics is 20,000 to 200,000 hertz. Dog whistles operate at 20,000 to 40,000 hertz. A bat emits a 30,000-hertz ultrasonic wave for its sonar.

Ultrasonic energy is often applied to liquids, solids and gases to produce desired effects and changes or to improve a process or a product. Ultrasonic energy is used to mix chemicals, photograph bones, homogenize milk, emulsify oils and clean metals.

The upper frequency limit for the propagation of ultrasonic sound waves is the thermal lattice vibration beyond which the material cannot follow the input sound wave. The smallest wavelength of sound is twice the interatomic distance, which for metal is about:

$$2 \times 10^{-10} \text{ meters}$$

which corresponds to a frequency of 1.25×10^{13} hertz.

This is the 21st harmonic of a 10-MHz quartz crystal. Wave periods are comparable with relaxation times at such high frequencies. High amplitude ultrasonic sound waves are also known as *sonic* sound waves. *Hypersound* waves refers to sound waves having frequencies greater than 1×10^{13} hertz.

Rayleigh surface waves propagate over a surface without affecting the bulk of the medium below the surface. *Rayleigh waves* are produced from unbalanced forces at the surface of a solid. They generate an elliptical motion of the medium whose amplitude decreases exponentially as the depth below the surface increases. *Ultrasonic* Rayleigh waves are propagated along the surface of a test object to detect flaws and cracks on or near the surface of the test object.

A *Lamb wave* is produced in a thin plate whose thickness is comparable to the wavelength of the Lamb wave. Lamb waves are complex waves. They move in asymmetrical or in symmetrical modes. Lamb waves are used to locate areas that are not bonded in laminated structures, to locate radial cracks in tubing and for quality control of plate and sheet stock.

TRANSDUCERS

There are gas-driven, liquid-driven and electromechanical ultrasonic transducers. Whistles and sirens are gas-driven transducers. Hydrodynamic oscillators and vibrating blade transducers are liquid-driven transducers. Piezoelectric and magnetostrictive transducers are electromechanical transducers. Ultrasonic transducers are classified according to the form of energy used to excite them into mechanical vibration and according to the medium into which the ultrasonic wave is propagated.

A *piezoelectric crystal* expands and contracts along the same axis where an alternating electric field is applied, as shown in *Figure 11-1*. The amplitude of the mechanical vibration of the crystal becomes very large when the frequency of the applied electric field approaches the natural frequency of any longitudinal mode of vibration of the crystal. Piezoelectric crystals are used to produce ultrasonic vibrations in the range of 20,000 to 200,000 hertz. Piezoelectric transducers are made from quartz, tourmaline (mineral occurring in various colors), Rochelle salt, ammonium dihydrogen phosphate, barium titanate, and ceramics with strong ferroelectric properties. Piezoelectric transducers provide stable ultrasound waves of narrow bandwidth. They can generate ultrasound waves over a wide range of frequencies. Some piezoelectric materials are hygroscopic and are therefore unable to sustain high power densities without fracturing. Some of the hygroscopic piezoelectric materials are also unstable.

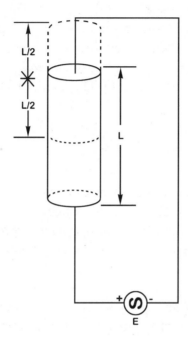

Figure 11-1. Piezoelectric transducer.

A rod made of a magnetizable material changes length when it is exposed to a varying magnetic field. The rod vibrates longitudinally when an alternating current flows through a coil surrounding the rod. The amplitude of the forced vibration of the rod becomes very large when the frequency of the applied current is the same as one of the normal longitudinal modes of vibration of the rod. Ultrasonic waves of this frequency are radiated. *Magnetostrictive transducers* are made from alloys of iron, nickel and cobalt. Magnetostrictive transducers are mechanically rugged and can produce large acoustical power at an efficiency of about 60 percent. They cannot produce extremely high frequency ultrasonic waves because of the extreme length of rod that would be required and because of hysteresis and eddy-current conversion losses.

Electromagnetic transducers generate ultrasound waves from the movement of a coil which carries a varying voltage in a constant amplitude magnetic field. Electromagnetic transducers operate like loudspeakers.

The *quality factor* Q (or the quality of resonance) of a system determines the frequency response of the system. The frequency response bandwidth is wide when the Q of the system is low. The frequency response bandwidth is narrow when the Q of the system is high. The magnification of an ultrasonic transducer is approximately equal to the quality factor Q. The quality factor Q is seven for quartz and water.

ABSORPTION

Ultrasonic energy is absorbed by gases because of the heat conduction and viscosity of gases. There is a delay in reaching equilibrium between the translational, rotational and vibrational energies of molecules. The delay affects the absorption of ultrasonic energy.

Ultrasonic energy is absorbed by solids because of lattice imperfections, ferromagnetic properties, ferroelectric properties, electron-photon interactions, thermal effects, grain boundary losses, thermoelastic relaxation, structural relaxation, acoustoelectric effects and nuclear magnetic resonance.

Ultrasound waves can be propagated much further in water than in gases and solids. Ultrasound waves of higher frequencies can be propagated in water than in gases and solids. Ultrasonic waves are not highly attenuated or absorbed in water.

APPLICATIONS

Ultrasonic waves are used to detect flaws, to improve processes and to control, monitor and measure the mechanical, physical, chemical and metallurgical properties of materials. The testing methods using ultrasonic waves are nondestructive. A transducer converts ultrasonic energy into a high frequency mechanical vibration of the medium through a coupling element, such as a horn.

Ultrasonics are used in industry for solidification, precipitation, agglomeration, emulsification, dispersion and cleaning of metals. They are also used to mix chemicals, homogenize milk and emulsify oils. Different types of ultrasonic generators and detectors are used.

Ultrasonics are used in medicine for detecting tumors and photographing bones. They are also used in biological measurements and diagnostic work.

Ultrasonics are used underwater to measure water depth in the mapping of the ocean floor. They are also used to detect submerged objects such as fish, mines and submarines.

Ultrasonics are also used in traffic control, fabric cleaning, aging of wines, packing of cement, counting, sorting and fog dispersion.

Problems

11-1.   Define ultrasonics.

11-2.   What limits the upper frequency for the propagation of ultrasonic sound waves?

11-3.   What can Rayleigh waves be used for?

11-4.   What can Lamb waves be used for?

11-5.   Name three types of ultrasonic transducers.

11-6.   An X-cut quartz crystal has a thickness of one millimeter and it is vibrating at resonance. Find its fundamental frequency:

where: Young's modulus for quartz is 7.9×10^{10} nt/m^2, and the density of quartz is 2650 kg/m^3.

11-7.   A magnetostrictive hydrophone is made from a nickel rod whose length is 0.2 meters. The rod is clamped at its center. What is the fundamental frequency of its longitudinal vibration if sound travels through the nickel rod at 4900 m/sec?

11-8.   Why is ultrasonic energy absorbed by gases?

11-9.   Why is ultrasound useful in water?

11-10. A delay time of one nanosecond is required by a computer to store information. If a copper wire with a diameter of 10^{-6} meters is used as an ultrasonic delay line, find the required length of the copper wire if sound travels through the copper wire at 3700 m/sec.

Part Two: Projects

Chapter 12

Design and Build World-Class Speaker Systems

Speaker systems are the last link in a home theater system. Audiophiles spend thousands of dollars on the latest compact disc player, AM/FM tuner, or surround-sound decoder and amplifier. These purchases are made only after studying their performance specifications. Speakers are often purchased based on price because the budget has been wiped out by the other purchases. A pair of good speaker systems can cost over one thousand dollars. Speaker designs and specifications can seem complicated.

Speaker systems are therefore usually the weakest link of a home theater system. There seems to be an infinite number of shapes and sizes of speaker systems on the market. Is it possible that many of the manufacturers are simply using "smoke and mirrors" to sell their systems as the best speaker systems available? This chapter teaches the audiophile how to design and build "first-class" speaker systems. The author has built dozens of speaker systems during the last thirty years for commercial and residential use.

SPEAKERS

Speakers are transducers that convert electrical energy into sound waves. There are four types of speakers: *full-range* speakers, which cover the entire audio frequency range of 20 Hz to 20 kHz; woofers, which cover the frequency range of 20 Hz to 3 kHz, or bass frequencies; *mid-range* speakers, which cover the frequency range of 1 kHz to 10 kHz; and finally, *tweeters*, which cover the frequency range of 4 kHz to 20 kHz, or upper audio frequencies.

The full-range speaker is usually a compromise design. The full-range speaker has a limited frequency response and it is usually used in compact and inexpensive "bookshelf" speaker systems. Woofers are usually eight to 15 inches in diameter. Felted paper or polypropylene is used to make woofer speaker cones. Mid-range speakers are usually three to eight inches in diameter. Paper, cloth or polypropylene is used to make mid-range speaker cones. Some mid-range speakers use a plastic dome to increase sound wave dispersion. Tweeters are usually two inches or less in diameter. Tweeter speaker cones are made of paper or cloth. Tweeters can make sound using a plastic or metal dome. They can also make sound using a piezoelectric element and diaphragm. Tweeters are very directional and they are often fitted with horns, baffles and mechanical "lenses" to help disperse the sound waves. Dome tweeters disperse sound waves more than cone tweeters.

Horns are designed to achieve different patterns of sound distribution. Horns act like acoustic transformers because they couple a high impedance at the throat of the horn to a low impedance at the mouth of the horn. Horns usually increase the electroacoustic efficiency of the drivers and they provide a better reproduction of sound.

Three types of horns are the *conical*, *exponential* and *hyperbolic*, and they are shown in *Figure 12-1*. The cross-sectional area of the conical horn expands the quickest while the cross-sectional area of the hyperbolic horn expands the slowest.

When a driver radiates sound waves into the throat of a horn, the air pressure is high and the particle-velocity is slow. When the sound waves expand into the mouth of the horn, the air pressure is low and the particle-velocity radiation is high.

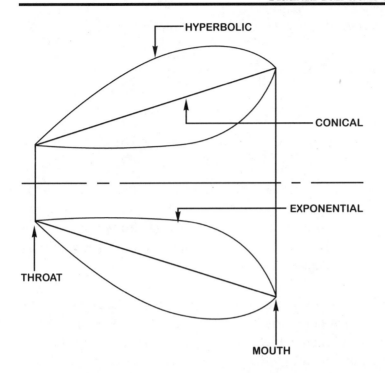

Figure 12-1.
Horns

The hyperbolic horn reaches the desired throat impedance the quickest of the three horns. It is therefore the most efficient horn down to its cutoff frequency. Non-linearity occurs inside the hyperbolic horn because high pressure develops when the expansion of the cross-sectional area of the horn is slow. The higher the hyperbolic horn is driven above its cutoff frequency, the greater the harmonic distortion. The harmonic distortion is generated by the non-linearity of the air in the hyperbolic horn.

The conical horn increases its cross-sectional area the fastest and it therefore generates the lowest distortion at higher frequencies. Unfortunately, the conical horn has a cutoff frequency that is about twenty times that of a hyperbolic horn.

The exponential horn has an expansion rate that is quick enough to keep the air pressure reasonably low. Distortion is therefore acceptably low. The cutoff frequency of an exponential horn is only about three times that of a hyperbolic horn. The exponential horn is the preferred choice for speaker design engineers.

The *differential equation of motion* for plane acoustic waves in a horn is:

$$\frac{d^2u}{dt^2} = \frac{c^2d^2u}{dx^2} + \frac{c^2dAdu}{Adx^2}$$

and its solution is:

$$u(x, t) = e^{-ax} [Ce^{j(wt - Bx)} + De^{j(wt + Bx)}]$$

where: u is displacement along the x-axis, a = m/2, with m being the flare constant of the horn, c is the speed of sound in air, k = w/c is the wave number, B = SQRT(k^2 - $0.25m^2$), and C and D are arbitrary constants.

The *radiating efficiency* of a horn is the ratio of the actual acoustic power radiated out of the horn to the acoustic power radiated by the same diaphragm which moves at the same velocity onto a cylindrical tube of infinite length and having the same cross-sectional area as the throat of the horn. The radiating efficiency of a horn is also called the transmission coefficient of the horn. The *transmission coefficient* of an exponential horn is:

$$T = \frac{1}{SQRT(1-[F_c/F]^2)}$$

where: F is the cutoff frequency of the horn and F is the frequency of the sound wave.

EXAMPLE 12-1

An exponential horn has a cutoff frequency of 1000 hertz. What is its transmission coefficient at 5000 hertz?

$$T = \frac{1}{SQRT(1 - [1/5]^2)} = \frac{1}{SQRT(1 - 0.04)}$$

$$= \frac{1}{SQRT(0.96)} = \frac{1}{0.98}$$

$$= 1.02$$

The *cutoff frequency* of a horn is the minimum frequency at which there is no propagation of sound waves inside the horn. The cutoff frequency of an exponential horn is:

$$F_c = \frac{mc}{2PI}$$

> where: m is the flare constant of the horn, c is the speed of sound in air, and PI = 3.1416.

The cutoff frequency of an *infinite exponential horn* is:

$$F_c = \frac{mc}{2PI}$$

A *multicellular array* is a group of horns driven by a common source. Each horn acts like a separate radiator of sound waves. An *acoustic lens* is a horn designed to control the directional spread of sound. Different arrays of obstacles are built into an acoustic lens. A *diffraction horn* approximates a point source because it is a narrow horn which expands uniformly in the vertical direction and because it is unflared in the horizontal direction.

A woofer and a mid-range speaker or a mid-range speaker and a tweeter can be combined into one unit or *coaxial* speaker. A coaxial speaker has two speaker cones and voice coils. A coaxial speaker has only one magnet.

A woofer, mid-range speaker and a tweeter can be combined into one unit or *triaxial* speaker. A triaxial speaker has three speaker cones and voice coils. A triaxial speaker has only one magnet.

Coaxial and triaxial speakers have wider frequency responses than a full-range speaker. Coaxial and triaxial speakers are about the same size as a full-range speaker.

Some woofers and tweeters have a *whizzer cone* to marginally improve the high frequency response of the speaker. The whizzer cone is attached to the cone and voice coil of the speaker. The whizzer cone has no voice coil.

There is some overlap between the various types of speakers, therefore, two-way or even three-way speaker systems may be designed. Of course one or more full-range speaker(s) may be used in a speaker system. If only one speaker is used in a speaker system, it should not be placed in the center of the speaker panel. Standing waves may be generated if the speaker is mounted in the center of the speaker panel.

There are several speaker specifications that should be understood when purchasing or building speaker systems. *Sound dispersion* is the spreading of sound waves as they leave the speaker.

At low frequencies the dispersion is *omnidirectional* or coming from all directions. The listener can therefore hear the sound waves from all directions. At high frequencies the dispersion is narrow or directional. The listener can only hear the sound waves that emanate from the front of the speaker.

Sound travels 13,504 inches per second. When the wavelength of the sound wave equals the diameter of the speaker cone, the dispersion changes from omnidirectional dispersion to directional dispersion. The approximate *omnidirectional frequency limit* is:

$$OFL = \frac{13,504}{D}$$

where: D is the diameter of the speaker cone in inches.

The speaker materials, cone geometry and speaker enclosure have an effect on the omnidirectional frequency limit.

EXAMPLE 12-2

What is the approximate omnidirectional frequency limit for a fifteen inch woofer?

$$OFL = \frac{13,504}{15}$$

$$= 900 \text{ Hz}$$

Speakers must be *damped* to prevent vibration at frequencies that are not in the input signal. Speakers are made with stiff cones to prevent unwanted vibrations. A large and heavy magnet is used to help damp the speaker. A large and heavy magnet also improves the low frequency response of the speaker. The speaker suspension can be stiff or pliable. The speaker cabinet design must complement the speaker's suspension to produce a clear and natural sound.

Harmonic and intermodulation distortion are characterized by frequencies in the speaker output sound waves which are not present in the speaker input signal. A proper speaker enclosure design can significantly reduce harmonic and intermodulation distortion.

Transient response of a speaker is the time it takes for a speaker cone to begin, from rest, to respond to a sudden electrical pulse. The stiffness of the speaker cone, suspension, and speaker cabinet affect the transient response of a speaker. The stiffer the speaker cone and suspension, the slower the speaker's transient response, because a stiff cone and suspension resist movement.

The *peak power rating* of a speaker is how much power the speaker can handle for only a very short time. The *average power rating* of a speaker is how much power the speaker can handle continuously. The speaker system must have an average power rating which is equal to or greater than the per-channel RMS power output of the amplifier that is used to drive the speaker system.

A speaker with a high power rating has a large voice coil because it can dissipate more heat. A large magnet enables the speaker cone to move large distances to produce high volume sounds at low frequencies. A woofer must therefore have a large magnet to produce high volume sounds at low frequencies.

Impedance is a measure of electrical resistance, in ohms, to AC signals. The speaker impedance should be the same as the output impedance of the driver amplifier. *Resistance* is a measure of the electrical resistance, in ohms, to DC signals. Speaker impedance is in the range of four to eight ohms. Resistance is five or six ohms.

Cone resonance or free-air resonance is the frequency at which a speaker resonates in free space. Below the frequency of cone resonance, acoustic output from the speaker decreases. Free-air resonance is determined by the mass of the speaker cone, other moving parts, and the compliance of the speaker's suspension. The resonance of a speaker increases when it is placed in an enclosure. It is an important specification for designing a speaker enclosure.

A *crossover network* is required in all systems using woofers, midranges, and tweeters, or woofers and tweeters. A crossover network splits the incoming audio signal into appropriate frequency ranges and feeds them to the proper speakers.

TYPES OF SPEAKER ENCLOSURES

A speaker that is not mounted on or in an enclosure does not operate efficiently, and therefore does not sound very good. All speakers are designed to be mounted inside an enclosure or cabinet.

The shape, size and construction of a speaker enclosure affects the overall performance of the speaker(s). The enclosure directs the sound waves, determines the frequency response of the system and controls the sound intensity. An enclosure with a rear panel stops front-to-end cancellation of sound waves. The rear panel improves the frequency response of the speaker system. A speaker cabinet with a rear panel increases the stiffness of the suspension system of the speaker cone by:

$$S = \frac{pc^2 A^2}{V} \text{ nt/m}$$

where: p is the air density in kg/m^3, c is the speed of sound in m/sec, A is the piston area in meters^2, and V is the volume of the cabinet in meters^3.

Sound waves are emitted from the front and the back of a speaker cone. A cabinet prevents the rear sound wave from interacting with the front sound wave, and thus prevents cancellation of the low frequency sound wave. Speaker performance is therefore vastly improved.

The *ported reflex* enclosure and the *sealed* enclosure are the two main types of speaker enclosures.

The sealed or acoustic-suspension enclosure is an airtight cabinet. The speaker cone acts like a piston because it compresses and de-compresses the air inside the sealed enclosure, as shown in *Figure 12-2*. This decreases the compliance of the speaker. Hence, a low compliance woofer (rolled polyfoam or butyl rubber suspension) must be used in a sealed enclosure. A sealed speaker system re-quires a lot of audio power to be driven adequately because they are inefficient. The acoustic-suspension enclosure increases the speaker's resonant frequency because it decreases its compliance. An acoustic-suspension enclosure improves the low frequency re-sponse of the speaker relative to ported reflex enclosures. The ported reflex design is more efficient than an acoustic-suspension design. If the sealed enclosure is very large, it is often called an *infinite baffle*.

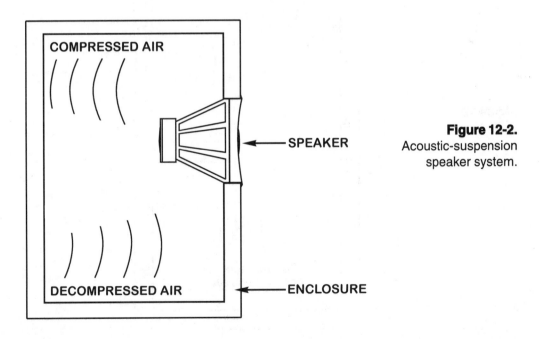

COMPRESSED AIR

SPEAKER

DECOMPRESSED AIR

ENCLOSURE

Figure 12-2.
Acoustic-suspension
speaker system.

The ported speaker system uses a port to reinforce the low frequencies. The *port* is a partial vent for the compressed and decompressed air, as shown in *Figure 12-3*. The port increases the effective enclosure volume; therefore, the ported system is efficient and can be driven properly by a "modest" amount of audio power. The ported speaker system can produce lower frequencies than an acoustic-suspension speaker system. A high compliance woofer (folded paper suspension) must be used in a ported speaker system. The ported speaker system is also known as a *ducted port*, as a *bass reflex*, or as a *Helmholtz resonator*.

The sealed speaker system is easier to design and build than the ported speaker system. However, this chapter explains how to design and build both types of speaker enclosures.

Figure 12-3.
Ported reflex speaker system.

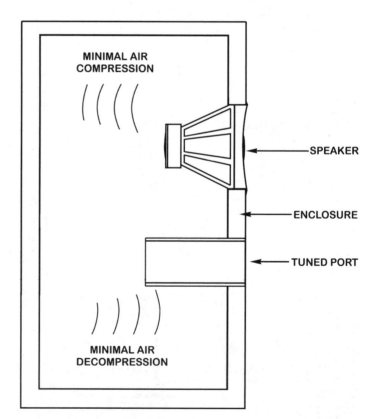

CABINET DESIGN

There are only a few simple rules to follow when designing the speaker enclosure. The diameter of the woofer (or of the full-range speaker in one-speaker systems) determines the internal volume of the speaker cabinet. *Table 12-1* is a list of the volumes required for the different woofer diameters.

The proportions of the speaker enclosure are critical. Cubes and very long or extremely shallow cabinets should be avoided to prevent intermodulation distortion. The internal cabinet height, width, and depth should be as close as possible to the ratio of **8:5:3**, respectively.

The port (or hole) and the duct (or tube) in a ported reflex speaker system allows the use of a relatively small enclosure while maintaining good low frequency response. When the enclosure is properly tuned, the bass response of the speaker is accentuated because the air vibrating in the port is in phase with the low frequencies of the woofer.

Woofer Diameter (inches)	Internal Cabinet Volume (cubic inches)
four	450 - 675
six	600 -1000
eight	1500 - 2500
ten	2500 - 5000
twelve	5000 - 10,000
fifteen	8000 - 15,000

Table 12-1.
Volume required for different woofer diameters.

If a ported speaker system is being built, use *Table 12-2* to find the duct-tube length or port area for a given volume and woofer free-air resonant frequency combination. For correct tuning, the ends of the duct tube must be clear from all obstructions (cabinet wall, bracing, etc.) for a distance at least equal to the diameter of the duct. In *Table 12-2*, the port area is in square inches and is calculated for three-quarter inch thick wood. D2 is a duct tube with a two-inch inside diameter, D3 is a duct tube with a three-inch inside diameter, and D5 is a duct tube with a four and three-quarter inch inside diameter.

EXAMPLE 12-3

Design an acoustic-suspension speaker system using a 12-inch mid-range speaker and a tweeter.

According to *Table 12-1*, a 12-inch woofer requires a speaker cabinet with a volume of 5000 to 10,000 cubic inches. The ratio of the inside height, inside width and inside depth of the speaker cabinet is 8:5:3. For a speaker cabinet with a volume of 7680 cubic inches, the inside height should be 32 inches, the inside width should be 20 inches and the inside depth should be 12 inches.

EXAMPLE 12-4

Design a ported reflex speaker system using an 8-inch woofer and a tweeter. The woofer has a free-air resonant frequency of 50 hertz.

According to *Table 12-1*, an 8-inch woofer requires a cabinet with a volume of 1500 to 2500 cubic inches. The ratio of the inside height, inside width and inside depth must be 8:5:3. For a speaker cabinet volume of 3240 cubic inches, the inside height is 24 inches, the inside width is 15 inches and the inside depth is 9 inches.

Because the woofer has a free-air resonant frequency of 50 Hz, a D5 (4-3/4" inside diameter) port tube should have a length of about 6.25". The port tube length is obtained by extrapolating the port tube lengths for cabinet volumes of 3000 and 4000 cubic inches.

Cabinet Volume in Cubic Inches					
f	1000	1500	2000	2500	3000	4000
25	closed	closed	D2 - 9.375	D2 - 7.625	D2 - 6	D2 - 4.125
30	closed	D2 - 9.125	D2 - 6.375	D2 - 4.75	D3 - 9.5	D3 - 6.5
35	D2 - 10.35	D2 - 6.25	D2 - 4.25	D3 - 8.125	D3 -6.375	D3 - 4.125
40	D2 - 7.5	D2 -4.375	D3 - 7.75	D3 - 5.625	D3 - 4.25	D5 - 8.75
45	D2 - 5.5	D2 - 3.125	D3 - 5.5	D3 - 4	D5 - 9.5	D5 - 6.125
50	D2 - 4.125	D3 - 6.25	D3 - 4	D5 - 9.125	D5 - 7	D5 - 4.125
55	D2 - 3.125	D3 - 4.625	D3 - 3	D5 - 7	D5 - 5	D5 - 2.75
60	D3 - 6.5	D3 - 3.5	D5 - 7.5	D5 - 5.125	D5 - 3.625	D5 - 1.75
65	D3 - 5.125	D3 - 2.5	D5 - 5.625	D5 - 3.75	D5 - 2.5	17 sq.in.
70	D3 - 4.125	D5 - 7.125	D5- 4.375	D5 - 2.625	D5 - 1.5	22 sq.in.
75	D3 - 3.25	D5 - 5.75	D5 - 3.25	D5 - 1.75	17 sq.in.	28 sq.in.
f	5000	6000	8000	10,000	12,500	15,000
25	D3 - 8	D3 - 6.25	D3 - 4	D5 - 9.125	D5 - 6.5	D5 - 4.75
30	D3 - 4.75	D3 - 3.5	D5 - 7.375	D5 - 5.125	D5 - 3.25	D5 - 2
35	D5 - 9.375	D5 - 7.125	D5 - 4.375	D5 - 2.625	15 sq.in.	20 sq.in
40	D5 - 6.25	D5 - 4.5	D5 - 2.375	15 sq.in	23 sq.in.	32 sq.in.
45	D5 - 4	D5 - 2.75	16 sq.in.	24 sq.in.	35 sq.in.	49 sq.in.
50	D5 - 2.5	D5 - 1.5	23 sq.in.	34 sq.in.	52 sq.in.	72 sq.in.
55	D5 - 1.5	20 sq.in.	32 sq.in.	48 sq.in.	72 sq.in.	closed
60	19 sq.in.	27 sq.in.	44 sq.in.	66 sq.in.	closed	closed
65	25 sq.in.	35 sq.in.	60 sq.in.	90 sq.in.	closed	closed
70	33 sq.in.	46 sq.in.	78 sq.in.	closed	closed	closed
75	42 sq.in.	60 sq.in.	closed	closed	closed	closed

Table 12-2. Port length for given volume and resonant frequency.

Cabinet Construction

The speaker cabinet should be constructed as rigidly and as heavily as possible out of particle board or plywood in order to minimize the acoustical energy that is radiated by the cabinet walls. Particle board and plywood do not resonate. Particle board or plywood that is three-quarters of an inch thick is adequate for speaker systems of about two cubic feet or less in volume. For larger speaker enclosures, two sheets of particle board or plywood should be laminated together to obtain a wall thickness of one and one-half inches. All joints must be made airtight by caulking them. Any resonant panels should be braced with cleats screwed to the inner surface. The interior (not the baffle) should be lined with two-inch thick fiberglass insulation. Insulation increases the effective volume of the enclosure. The insulation should be fastened to the interior of the cabinet with staples or large-head tacks such as roofing nails. A curtain of insulation should be hung down the full width and height of the enclosure just behind the loudspeakers. However, do not obscure the duct opening.

The loudspeakers should be mounted on the outside of the baffle and flush with the baffle surface. This ensures that none of the sound waves emanating from the speakers are blocked by the baffle.

Wiring Speaker Systems

Speakers are polarized. Either a "+" symbol or a red dot is used to mark the positive speaker terminal. If the terminals are not marked, the speaker polarity may be determined with a "D" size flashlight battery. Momentarily connect the battery to the speaker terminals and watch the direction of the speaker cone movement. If the cone moves outward, the positive terminal of the battery is connected to the positive speaker terminal. The negative terminal of the battery is connected to the negative speaker terminal. If the cone moves inward, the negative terminal of the battery is connected to the positive speaker terminal. The positive terminal of the battery is connected to the negative speaker terminal.

A crossover network divides the input signal into two or more output signals. A two-way crossover network has two outputs— one for a woofer and one for a tweeter. A three-way crossover network has three outputs—one for a woofer, one for a mid-range speaker and one for a tweeter.

A two-way crossover network has one *crossover frequency*. The crossover frequency must be within the frequency response curves of the woofer and the tweeter. A two-way crossover network consists of a passive low-pass filter and a passive high-pass filter, as shown in *Figure 12-4*. The output of the low-pass filter is connected to the woofer. The output of the high-pass filter is connected to the tweeter. The frequency response of the two-way crossover network is also shown in *Figure 12-4*.

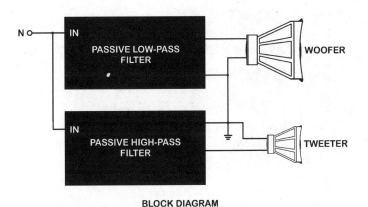

BLOCK DIAGRAM

Figure 12-4.
Two-way crossover network.

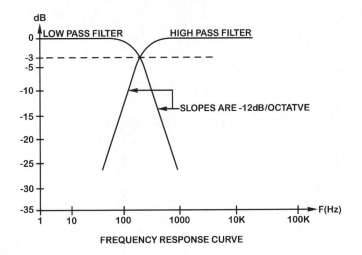

FREQUENCY RESPONSE CURVE

A three-way crossover network has two crossover frequencies. The lower crossover frequency must be within the frequency response curves of the woofer and the mid-range speaker. The higher crossover frequency must be within the frequency response curves of the mid-range speaker and the tweeter. A three-way crossover network consists of a passive low-pass filter, a passive bandpass filter and a passive high-pass filter, as shown in *Figure 12-5*. The woofer is connected to the output of the low-pass filter. The mid-range speaker is connected to the output of the bandpass filter. The tweeter is connected to the output of the high-pass filter. The frequency response of a three-way crossover network is also shown in *Figure 12-5*.

Figure 12-5.
Three-way crossover network.

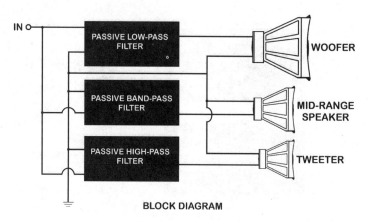

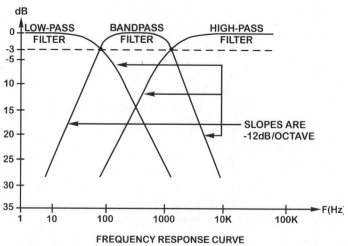

The ideal slopes of the passive filter responses is 12 decibels per octave. If the slope is less than 12 decibels per octave, the tweeter is not adequately protected from low frequency signals. If the slope is greater than 12 decibels per octave, transient distortion is increased in the woofer and in the tweeter.

An L-pad is used to alter the response of a speaker system by controlling the amplitude of the electrical signal that is fed to the speaker. The L-pad is connected either between the amplifier and speaker or between the crossover network and speaker. L-pads will burn out if their maximum power rating is exceeded.

The positive terminal of the speaker system must be connected to the positive terminal of the amplifier. The negative terminal of the speaker system must be connected to the negative terminal of the amplifier. The interconnecting wires must be at least 18 AWG (American Wire Gauge) to carry 100 watts or less of audio power distances of 50 feet or less.

A fuse should be installed in series with the positive terminal wire of the speaker system, as shown in *Figure 12-6*. A second fuse can also be used to protect the tweeter. The fuse rating is:

$$I = SQRT(\frac{P}{Z})$$

where: P is the peak or maximum power handling rating of the speaker, and Z is the speaker impedance.

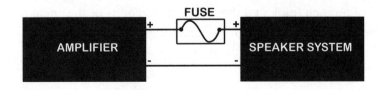

Figure 12-6.
Fuse protection for a speaker system.

EXAMPLE 12-5

Calculate the required fuse rating of an eight-ohm speaker with a peak power rating of 50 watts.

$$I = SQRT(\frac{50}{8})$$
$$= SQRT(6.25)$$
$$= 2.5 \text{ amperes}$$

TWO SPEAKER SYSTEMS

The centerpieces of the author's audio/visual system are two speaker systems with a volume of four and one-half cubic feet each. Each speaker system has a 15-inch woofer, a dual-cone eight-inch mid-range speaker, and a six-inch exponential horn tweeter. The three-way crossover network operates at -12 dB/octave at about 800 Hz and at about 5 kHz. The side walls as well as the tops and bottoms are two inches thick, which was obtained by laminating two three-quarter inch thick sheets with a one-half inch thick sheet of particle board, covered with a thin layer of mahogany veneer. The fronts and backs are one and one-half inches thick. Each speaker system weighs about 200 pounds. The inside dimensions are 32 inches high, 20 inches wide, and 12 inches deep. The woofer and speaker enclosure are designed to resonate at 30 Hz.

The author has also designed a pair of "book shelf" speaker systems. Each system contains a four-inch long-throw woofer and a two and one-half inch tweeter. The crossover network has a crossover frequency of about 2,500 Hz. The cabinet walls are three-quarters of an inch thick and the inside dimensions of the speaker enclosures are ten and five-eighths inches high, six and five-eighths inches wide, and four inches deep.

SPEAKER SYSTEM PLACEMENT

Stereo speaker systems must be placed at least eight feet apart. They may be placed closer together in very small rooms. If the speaker cabinets are too close together, the spatial quality of the sound is sacrificed.

The ideal listening distance from the speakers is between one and two times the distance by which the speaker systems are separated. If the speaker systems are 10 feet apart, then the ideal listening distance is 10 to 20 feet in front of the speaker systems. Listening too close to two widely separated speaker systems gives the impression of two distinctly separate sound sources instead of the desired broad front of sound.

A wall reflects sound waves. At low frequencies, the reflected sound waves are in phase with the incident sound waves. If the speaker system is placed near a wall, the wall doubles the low frequency response of the speaker system. If the speaker system is placed near a corner, the second wall again doubles the low frequency response of the speaker system. Placing the speaker system on the floor once again doubles the low frequency response of the speaker system. The front-channel speaker systems should be placed in two corners of the listening room because the human ear is relatively insensitive to low frequency sounds.

CONCLUSION

Excellent speaker systems can now be designed by anyone with minimal carpentry skills. The speakers to be used should be selected first and then the enclosures can be designed using the procedures outlined in this article. The speaker systems no longer have to be the weakest link of a stereo system.

Chapter 13

Nine-Band
Graphic Equalizer

An equalizer is a sophisticated tone control. The audio signal is divided into nine one-octave bands. The gain of each band can be adjusted to correct for room acoustics or for loudspeaker characteristics. For example, if the room has a resonant frequency, an equalizer can be used to effectively decrease the output at the resonant frequency of the room without affecting the output of the rest of the audio signal. The equalizer can also be used to add special effects, to improve speech intelligibility, and to remove the vocal track from a recording so that the user can sing along while the musicians play.

The room equalizer separates the audio signal into nine one-octave bands. The gain of each band filter can be adjusted to correct for listening room acoustics. The nine bands are summed together to reconstitute the equalized audio signal.

The equalizer has a flat frequency response when the gain controls are set to their mid-positions. Each band filter has a sharp roll-off curve of twelve decibels per octave. Adjusting the gain of one band filter will not affect the gain of the adjacent band filters. The nine band filters allow flexibility without making adjustments difficult or tedious. The equalizer does not distort the audio signal and it does not introduce hum or noise to the audio signal.

CIRCUIT DESCRIPTION

The nine-band graphic equalizer consists of one unity-gain input buffer amplifier, nine variable gain amplifiers, nine band filters and four summing amplifiers, as shown in *Figure 13-1*, a block diagram of the equalizer. Two equalizers are required for a stereo system.

The circled letters and integrated circuits (U1-U30) in *Figure 13-1* correspond to the circled letters and integrated circuits in *Figures 13-2, 13-3* and *13-4*.

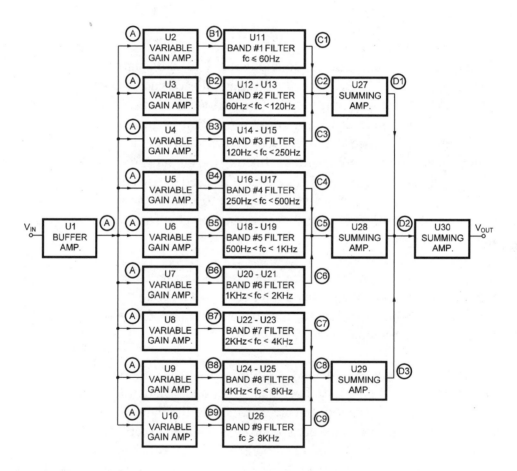

Figure 13-1.
Block diagram
of the equalizer.

The unity-gain input buffer amplifier has a high input impedance and a low output impedance. The input audio signal is therefore not loaded by the equalizer. The schematic of the input buffer amplifier is shown in *Figure 13-2*. The buffer amplifier is a non-inverting amplifier. The DC component of the audio signal is blocked by coupling capacitor, C1. The output of the buffer amplifier is connected to the inputs of the nine variable-gain amplifiers, as shown in *Figure 13-1*.

The variable-gain amplifier is a non-inverting amplifier, as shown in *Figure 13-2*. Its gain is:

$$A_v = 1 + \frac{R1 + RP1A}{R2 + RP1B}$$

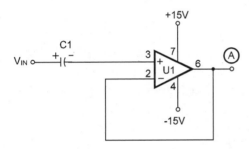

BUFFER AMPLIFIER (ONE REQUIRED)

$A_v = 1$

Figure 13-2.
Schematics of buffer
and variable gain
amplifier.

$$A_v = 1 + \frac{(R1 + RP1A)}{(R2 + RP1B)}$$

VARIABLE - GAIN AMPLIFIER (NINE REQUIRED)

$$1 + \frac{R1}{(RP1 + R2)} \leq A_v \leq 1 + \frac{(R1 + RP1)}{R2}$$

It can be adjusted from 1.17 to 6.0 with gain control RP1. When the gain control is set to its middle (reference) position, the gain of the variable-gain amplifier is 2. The gain of each variable-gain amplifier, with respect to its potentiometer's reference position, can be adjusted for a maximum gain of about 10 decibels, for a maximum attenuation of about six decibels or anywhere in between the two extremes. The output of each of the nine variable-gain amplifiers is connected to a band filter, shown in *Figure 13-1*.

The audio signal is split up into nine one-octave bands. The lowest octave band (band #1) requires a low-pass filter, the highest octave band (band #9) requires a high-pass filter, and the other one-octave bands require bandpass filters. The schematics of the low-pass, high-pass and bandpass filters are shown in *Figure 13-3*. The frequency response curves of the filters are also shown in *Figure 13-3*.

The low-pass filter is an equal-component unity-gain Sallen and Key filter. It is a second-order filter. The Sallen and Key filter has good sensitivity characteristics. The high-pass filter is also an equal-component unity-gain Sallen and Key filter. It is also a second-order filter.

The bandpass filter is an equal-component fourth-order Sallen and Key filter. It consists of a second-order Sallen and Key low-pass filter connected in series with a second-order Sallen and Key high-pass filter, as shown in *Figure 13-3*. The bandpass filter has a gain of less than unity (gain of one-third) because the cutoff frequencies of the low-pass and high-pass filters are only one octave apart, as indicated in the frequency response curve of the bandpass filter.

The filters all have 12-decibel-per-octave roll-off curves. The output of each band filter is connected to an input of a summing amplifier, per *Figure 13-1*.

Impedance and frequency scaling are used to calculate the frequency selective components of the filters. The three-step design procedure is:

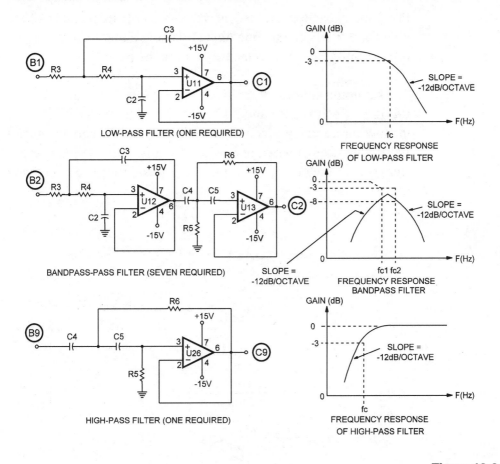

Figure 13-3.
Schematics of low-pass,
high-pass and bandpass filters.

1. Frequency scaling:

$$a = 2 \times PI \times f_c = 6.283\ f_c$$

where: f_c is the required cutoff frequency in hertz.

2. Arbitrarily select capacitor C.

3. Impedance scaling:

$$R = \frac{1}{aC}$$

where: R is the required resistor in ohms, and C is the selected capacitor in farads.

Precision capacitors are expensive. The cutoff frequencies of the nine band filters can be fine tuned by using either the next higher or next lower standard value resistors in the band filters.

The summing amplifiers are inverting amplifiers, as shown in *Figure 13-4*. Resistors R7 and R17 are selected to reduce the gain of the low-pass and high-pass filters to ensure that the equalizer has a flat frequency response curve when the nine variable-gain amplifier potentiometers are set to their mid-positions. The gain of the

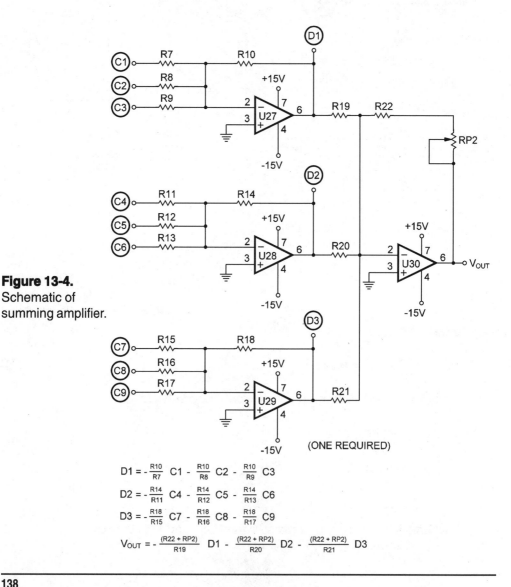

Figure 13-4.
Schematic of summing amplifier.

$$D1 = -\frac{R10}{R7}\,C1 - \frac{R10}{R8}\,C2 - \frac{R10}{R9}\,C3$$

$$D2 = -\frac{R14}{R11}\,C4 - \frac{R14}{R12}\,C5 - \frac{R14}{R13}\,C6$$

$$D3 = -\frac{R18}{R15}\,C7 - \frac{R18}{R16}\,C8 - \frac{R18}{R17}\,C9$$

$$V_{OUT} = -\frac{(R22 + RP2)}{R19}\,D1 - \frac{(R22 + RP2)}{R20}\,D2 - \frac{(R22 + RP2)}{R21}\,D3$$

bandpass filter is less than the gain of the low-pass and high-pass filters because the cutoff frequencies of the low-pass and high-pass filters are only one octave apart, as shown in *Figure 13-3*. The gain of the equalizer can be adjusted from unity to six by potentiometer RP2.

The power supply for the equalizer is shown in *Figure 13-5*. Switch S1 connects the alternating household line current to the primary winding of transformer T1. Fuse F1 protects the equalizer circuits. Transformer T1 steps down the alternating household line voltage to a lower alternating voltage. The diode bridge, which consists of diodes D1-D4, rectifies the alternating low voltage into a pulsating DC voltage. Capacitors C6 and C7 smooth out the pulsating DC voltage into a DC voltage with a ripple or AC component. The AC component of the DC voltage is reduced by positive voltage regulator U31 and negative voltage regulator U32. Capacitors C8

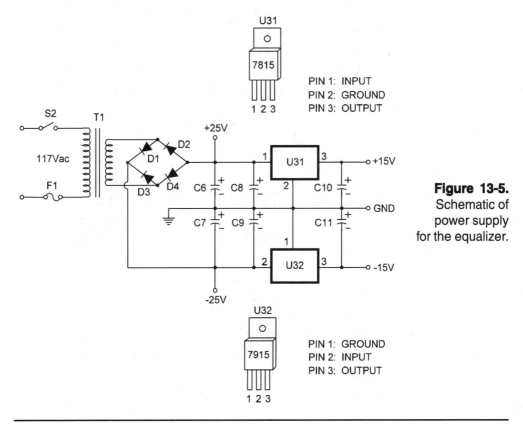

Figure 13-5. Schematic of power supply for the equalizer.

and C9 are required if the voltage regulators are located more than a few inches from filter capacitors C6 and C7. Capacitors C10 and C11 improve the transient responses of the positive and negative voltage regulators, respectively.

CONSTRUCTION AND TESTING

The equalizer may be built on one large piece of perf board or on a few small pieces of perf board. The equalizer is a complex project and it is recommended that the equalizer is assembled and tested in stages. *Tables 13-1* and *13-2* are the parts lists for the equalizer.

If the reader has little or no experience soldering, he or she should practice soldering some components on a scrap piece of perf board. Integrated circuit sockets should be used for integrated circuits U1-U30.

The power supply should be assembled and tested first. The schematic of the power supply is shown in *Figure 13-5*. No integrated circuit sockets are required for integrated circuits U31 and U32. Diodes D1-D4, capacitors C6-C11, and integrated circuits U31 and U32 must be properly oriented or the power supply will not function.

Measure the voltages across capacitors C6 and C7. They should be equal, of opposite polarities, and about 25 volts. If they are not, verify the connections of the transformer, the diodes, and capacitors C6-C9. The output voltages of U31 and U32 should be plus and minus 15 volts, respectively. If they are not, verify the connections of integrated circuits U31 and U32 and capacitors C10 and C11.

Once the power supply is working, the rest of the equalizer can be built and tested in sections. The first section of the equalizer to be built consists of the buffer amplifier and the nine variable-gain amplifiers. The schematics of the buffer and variable-gain amplifiers are shown in *Figure 13-2*. The output of the buffer amplifier goes to the inputs of all nine variable-gain amplifiers, as shown in

Figure 13-1. This section can be verified by feeding a one-volt peak-to-peak sinusoidal signal to the input of the buffer amplifier. There should be a sinusoidal signal at the output of the buffer amplifier and at the output of each variable-gain amplifier. The output of each variable-gain amplifier can be varied from one volt peak-to-peak to six volts peak-to-peak by adjusting its gain control RP1.

The nine-band filters shown in *Figure 13-3* can be built next. Band #1 filter is a low-pass filter. Band #9 filter is a high-pass filter. The other band filters are bandpass. They should be connected to the section of the equalizer already built, as shown in *Figure 13-1*. The operation of each filter can be tested by feeding a one-volt peak-to-peak sinusoidal signal to the input of the buffer amplifier.

Each band filter should pass only the frequencies indicated in *Figure 13-1* and *Table 13-2*. If one or more band filters are not functioning properly, verify their connections.

The equalizer can then be completed by building the summing amplifiers shown in *Figure 13-4*. The output at D1 of *Figure 13-4* should pass the frequencies of band filters 1, 2 and 3. The output at D2 should pass the frequencies of band filters 4, 5 and 6. The output at D3 should pass the frequencies of band filters 7, 8 and 9. The output of the equalizer should be a composite of the nine band filter outputs. If not, verify the connections of the appropriate summing amplifier.

The LM741 operational amplifier was chosen because it is available in all electronic parts stores and because it is inexpensive. Almost any operational amplifier may be used. The reader must be sure that he or she connects the proper pins for the operational amplifier used into the equalizer circuits.

The potentiometers may be rotary controls or slider controls. If slider potentiometers are used, slots must be cut into the front panel to accommodate them.

INSTALLATION AND USE

The input of the equalizer is connected to the output of a preamplifier. The output of the equalizer is connected to the input of a power amplifier. The equalizer may also be connected into the tape monitor loop of a receiver or of an integrated amplifier (preamplifier and power amplifier combined in one unit). Consult the owner's manual for the proper connections to the tape monitor loop of the receiver or of the integrated amplifier.

Set the nine variable-gain amplifier gain controls (RP1) to their mid-positions. Set the equalizer volume control (RP2) fully counterclockwise. Power up the audio system and the equalizer. Slowly rotate the equalizer volume control (RP2) for the desired listening volume. Adjust the nine gain controls (RP1) as necessary to achieve the desired sound affects. The equalizer should give years of trouble-free service.

Table 13-1.
Parts list for
the equalizer.

> All resistors are 1/4W at 5% unless otherwise noted.
>
> R1, R2, R8-R16, R18-R22: 10K
> R3, R4: see Table 13-2
> R5, R6: see Table 13-2
> R7, R17: 30K
> RP1: 50K linear potentiometer
> RP2: 50K audio taper potentiometer
> C1: 10 uF at 16V
> C2, C3: see Table 13-2
> C4, C5: see Table 13-2
> C6, C7: 1000 uF at 35V
> C8, C9: 0.33 uF at 35V tantalum
> C10, C11: 1.0 uF at 16V tantalum
> D1-D4:1N4004 diode
> U1-U30: LM741 operational amplifier
> U31: MC7815 positive voltage regulator
> U32: MC7915 negative voltage regulator
> T1: 36 VCT secondary at 1A step-down power transformer
> S1: SPST switch
> F1: 1A slow-blow fuse

Band No.	Freq. (Hz)	R3, R4 (ohms)	C2, C3 (uF)	R5, R6 (ohms)	C4, C5 (uF)
	All resistors are 1/4W at 1%.				
1	<60	27K	0.1	—	—
2	60-120	12K	0.1	27K	0.1
3	120-250	62K	0.01	12K	0.1
4	250-500	33K	0.01	62K	0.01
5	500-1000	15K	0.01	33K	0.01
6	1000-2000	82K	0.001	15K	0.01
7	2000-4000	39K	0.001	82K	0.001
8	4000-8000	18K	0.001	39K	0.001
9	>8000	—	—	18K	0.001

Table 13-2.
Parts list for
the band filters.

Chapter 14

Designing Active Crossover Networks

Each channel of a conventional audio system consists of a preamplifier, a power amplifier, and a speaker system. The block diagram of one channel of a conventional audio system is shown in *Figure 14-1*. The passive crossover network is located inside the speaker cabinet.

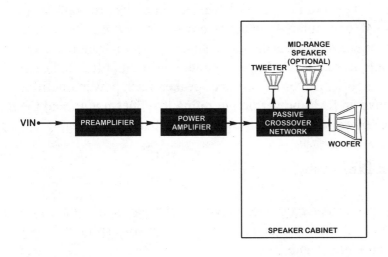

Figure 14-1.
Block diagram of a conventional audio system.

A *passive* crossover network requires inductors and capacitors. At frequencies lower than 1000 Hz, inductors are bulky and expensive. The dissipative losses of inductors increase with frequency. If an inductor is miniaturized, dissipative losses increase. Dissipative losses are power losses due to the wire windings of the inductor. Passive networks usually require low impedance signal sources. The performance of a passive network depends on the impedance of the load it is driving. Passive stages cannot be connected in series without buffer amplifiers. Passive components cannot provide gain. Passive crossover networks are therefore not well suited for audio circuits.

Active crossover networks do not require inductors. Most transfer functions can be emulated using only resistors and capacitors as feedback components. Resistors and capacitors are inexpensive components. The operational amplifier is an ideal active device because it has a high input impedance, a low output impedance and a high open-loop gain. Operational-amplifier stages can be connected in series without buffer stages. The frequency response curves of active circuits are independent of source and load impedances. Active circuits can provide gain or attenuation and they can be tuned as well.

An audio system can be designed to take advantage of an active crossover network. The block diagram of one channel of the audio system is shown in *Figure 14-2*. The output of the preamplifier is connected to the input of an active crossover network. The active crossover network splits the audio signal into two or more bands. Each output is connected to a separate power amplifier and each power amplifier drives at least one speaker. Each power amplifier must have an RMS power output rating that does not exceed the nominal power rating(s) of the speaker(s) that it is driving.

ACTIVE CROSSOVER NETWORKS

A one-way active crossover network splits its input signal into two bands. A two-way active crossover network splits its input signal into three bands.

A one-way active crossover network consists of an active low-pass filter and an active high-pass filter. The block diagram of a one-way active crossover network is shown in *Figure 14-3*. The cutoff frequencies of the two filters are equal. The output of the

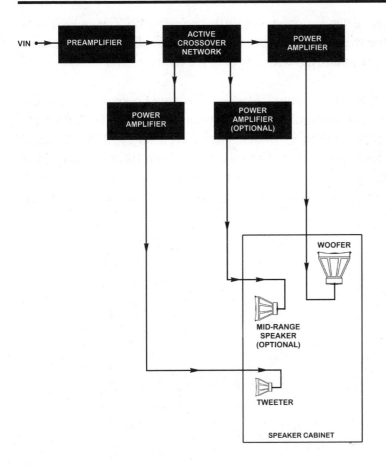

Figure 14-2.
Block diagram of
an audio system
with an active
crossover network.

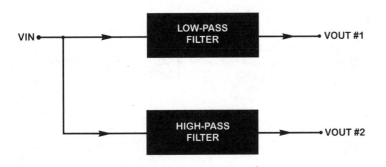

Figure 14-3.
Block diagram of a
one-way active
crossover network.

low-pass filter is connected to the power amplifier, which is to drive the woofer. The output of the high-pass filter is connected to the power amplifier, which is to drive the tweeter.

The schematic of a one-way active crossover network is shown in *Figure 14-4*. Both filters are unity-gain second-order Sallen and Key equal-component filters. They have sharp roll-off curves of 12 decibels per octave. The resistors of both filters are equal and the capacitors of both filters are equal. The frequency response curves of the one-way active crossover network are shown in *Figure 14-5*.

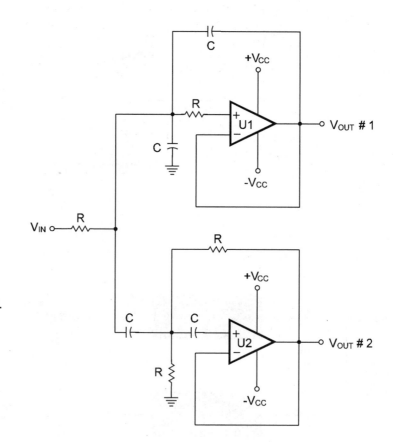

Figure 14-4.
Schematic of a one-way active crossover network.

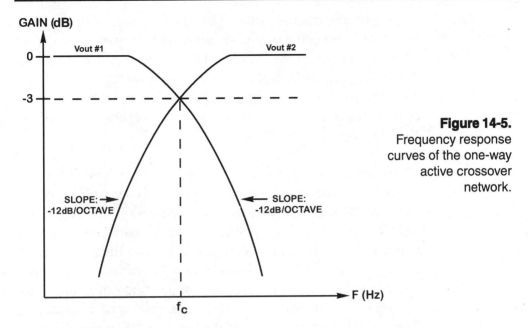

Figure 14-5.
Frequency response
curves of the one-way
active crossover
network.

A two-way active crossover network consists of an active low-pass filter, an active high-pass filter and an active bandpass filter. The active bandpass filter is an active low-pass filter connected in series with an active high-pass filter. The block diagram of a two-way active crossover network is shown in *Figure 14-6*. The cutoff frequency of the low-pass filter must equal the lower cutoff fre-

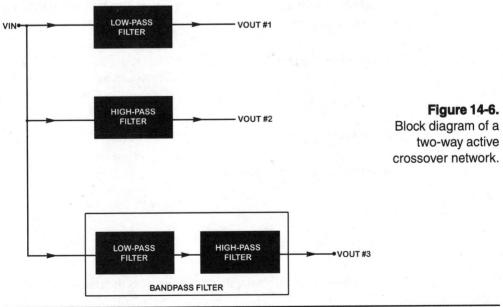

Figure 14-6.
Block diagram of a
two-way active
crossover network.

quency of the bandpass filter. The cutoff frequency of the high-pass filter must equal the upper cutoff frequency of the bandpass filter. The output of the low-pass filter is connected to the power amplifier, to drive the woofer. The output of the high-pass filter is connected to the power amplifier, to drive the tweeter. The output of the bandpass filter is connected to the power amplifier, to drive the mid-range speaker.

The schematic of a two-way active crossover network is shown in *Figure 14-7*. The three filters are unity-gain Sallen and Key equal-component filters. The low-pass and high-pass filters are second-order filters. The bandpass filter is a fourth-order, low-pass filter connected in series with a high-pass filter. The filters have sharp roll-off curves of 12 decibels per octave. The resistors and capacitors of the low-pass filter and the resistors and capacitors of the high-pass filter stage of the bandpass filter are equal, respectively. The resistors and capacitors of the high-pass filter and the resistors and capacitors of the low-pass filter stage of the bandpass filter are equal, respectively. The frequency response curves of the two-way active crossover network are shown in *Figure 14-8*.

DESIGNING ACTIVE CROSSOVER NETWORKS

The resistors and capacitors of each filter can be calculated by using frequency and impedance scaling. The three design steps are:

1. Frequency scaling:

$$a = 2 \times PI \times f_c = 6.283 \, f_c$$

where: f_c is the cutoff frequency in hertz.

2. Arbitrarily choose capacitor C.

3. Impedance scaling:

$$R = \frac{1}{aC}$$

where: C is the capacitor in farads, R is the resistor in ohms.

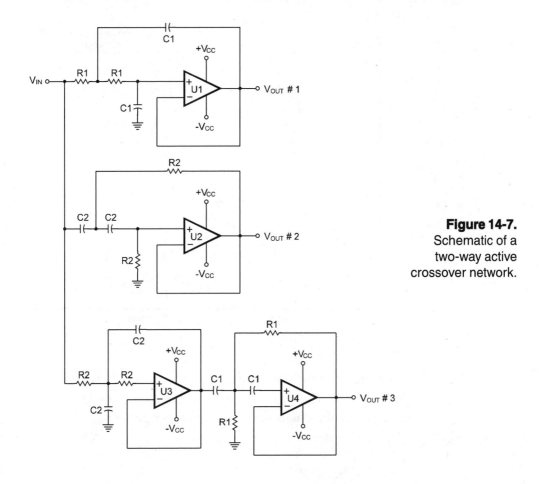

Figure 14-7.
Schematic of a
two-way active
crossover network.

EXAMPLE 14-1

Design a one-way active crossover network with a crossover frequency of 2000 Hz.

1. $a = 2 \times PI \times f_c = 2 \times PI \times 2000 = 12,566$

2. Let $C = 0.001$ uF

3. $R = \dfrac{1}{aC} = \dfrac{1}{(12,566 \times 0.001 \times 10^{-6})} = 79,580$ ohms.

(Using the nearest standard resistor, R = 82 kohms.)

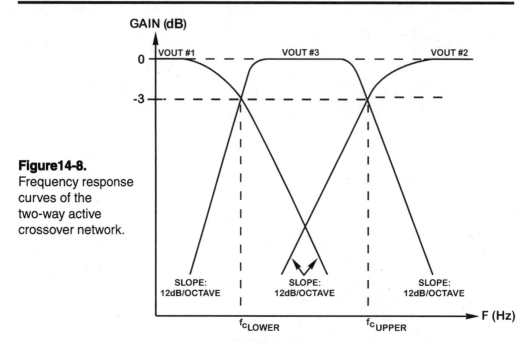

Figure14-8.
Frequency response curves of the two-way active crossover network.

The schematic of the one-way active crossover network is shown in *Figure 14-4*. The power supply shown in *Figure 13-5* can be used to power the one-way active crossover network.

EXAMPLE 14-2

Design a two-way active crossover network with a lower crossover frequency of 1000 Hz and an upper crossover frequency of 4000 Hz.

The schematic of a two-way crossover network is shown in *Figure 14-7*. Components R1 and C1 are calculated using the lower crossover frequency of 1000 Hz. Components R2 and C2 are calculated using the upper crossover frequency of 4000 Hz.

For the lower crossover frequency of 1000 Hz:

1. a1 = 2 x PI x f_c = 2 x PI x 1000 = 6283

2. Let C1 = 0.01 uF

3. $R1 = \dfrac{1}{a1C1} = \dfrac{1}{(6283 \times 0.01 \times 10^{-6})} = 15{,}916$ ohms

(using the nearest standard resistor, R1 = 15 kohms)

For the upper crossover frequency of 4000 Hz:

1. $a2 = 2 \times PI \times f_c = 2 \times PI \times 4000 = 25{,}132$

2. Let C2 = 0.001 uF

3. $R2 = \dfrac{1}{a2C2} = \dfrac{1}{(25{,}132 \times 0.001 \times 10^{-6})} = 39{,}790$ ohms

(using the nearest standard resistor, R2 = 39 kohms)

The power supply shown in *Figure 13-5* can be used to power the two-way active crossover network.

Choosing an Operational Amplifier

Important operational-amplifier specifications for audio applications are slew rate, unity-gain bandwidth, CMRR, PSRR and noise level.

Slew rate is the time required for the amplifier's output to respond to an input signal. Slew rate is critical in high speed and low distortion circuits.

Unity-gain bandwidth specifies the highest frequency that an amplifier will pass at a gain of unity without attenuation in the amplification process.

CMRR or *common-mode rejection ratio* is the ability of an operational amplifier to cancel out, within the device, common signals fed to its inverting and non-inverting inputs.

PSRR or *power supply rejection ratio* is the ability of an operational amplifier to prevent power supply fluctuations from showing up in the output signal.

Low noise is essential in high quality audio and video circuits. *Noise* is any unwanted signal that is present in the output signal of the operational amplifier.

The specifications of several operational amplifiers are listed in *Table 14-1*. The 4136 quad operational amplifier has a good compromise of the specifications listed in *Table 14-1*. The pin configuration of the 4136 quad operational amplifier is shown in *Figure 14-9*. The 4136 is a quad operational amplifier; therefore, more functions may be handled in a smaller area and at a lower cost than with single-amplifier integrated circuits.

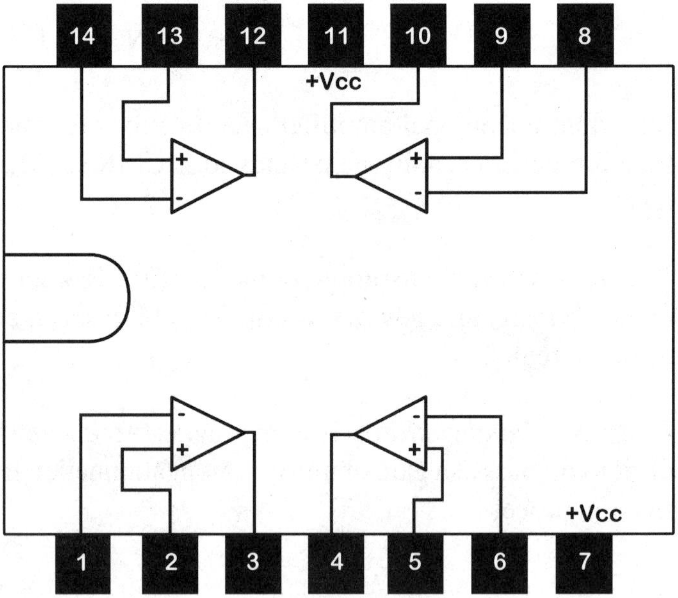

Figure 14-9.
Pin configuration
of the 4136 quad
operational amplifier.

	741	1458	324	308	712	5534	4136
Number of op-amps in IC	1	2	4	1	1	1	4
Input impedance (ohms)	2.2M	1M	4M	40M	40k	100k	5M
Slew rate (V/uS)	0.5	0.5	0.5	0.3	—	13	1.0
Unity-gain bandwidth (Hz)	1.5M	1M	1M	1M	5M	10M	3M
CMRR (db)	90	90	70	100	100	100	90
PSRR (db)	96	96	100	96	—	10	30uV/V
Offset voltage (mV)	2	1	2	2	—	0.5	10
Noise voltage (nV/SQRT(Hz))	—	—	—	60	—	4	10
Average quiescent current	1.7	3.0	1.5	0.3	—	4	7(mA)

Table 14-1.
Typical specifications for
operational amplifiers.

Chapter 15

Home Theater
Surround-Sound System

Surround sound is the reproduction of the spacious acoustics of a live performance in a residential listening room. Rear speakers and a decoder are required for surround sound.

David Hafler invented a four-channel system many years ago. An additional pair of series-connected speakers was connected directly across the positive output terminals of a stereo power amplifier. The pair of series-connected speakers served as the rear speakers of a four-channel system. The rear speakers received a signal consisting of the difference between the left and right channels. When a monophonic (the same) signal was fed to the left and right channels, the rear speakers were silent, as there was no difference signal.

Peter Scheiber invented a system of encoding the rear channel information on existing stereo channels. He used a decoder to decode the rear-channel signal and feed it to the rear-channel amplifier. His surround-sound decoder was designed for movie theaters.

Surround sound in the residential environment is possible because of the recent introduction of stereo television and stereo video cassette recorder. Most households have a stereo system where the left front speaker reproduces the left side of the performance and the right front speaker reproduces the right side of the performance. There are no rear speakers in a stereo system.

A home theater system is a surround-sound system. This home theater system consists of two preamplifiers, two sub-woofer simulators, a surround-sound decoder, a voice decoder, a rear channel power amplifier, a voice channel power amplifier and two front channel power amplifiers, as shown in *Figure 15-1*, which is a block diagram of the home theater surround-sound system.

Dolby Laboratories invented a system of encoding the rear channel information on the existing stereo channels. A decoder is required to decode the rear channel signal and to feed it to the rear channel power amplifier. Most surround-sound decoders are designed for use in the movie theater. This home theater system was designed by the author for use in residential environments.

The most popular surround-sound system is the Dolby System. The rear channel signal is generated by subtracting the right channel signal from the left channel signal. Therefore, a decoder (L - R) is required for the home theater system. The decoder passes the left input signal in phase and it inverts the right input signal. High frequencies are attenuated by a low-pass filter. The low-pass filter passes all frequencies in its passband in phase with its input signal. The rear channel amplitude is varied by the rear channel gain control. The rear channel signal is amplified by the rear power amplifier and sent to the rear channel speakers.

The average residential listening room is about 15 feet wide. The left and right speakers are placed in the front corners of the listening room. Speech should appear to emanate from the television. A center voice channel (L + R) is often required in a home theater system because the distance between the left and right speakers can create an audio "hole." A bandpass filter passes only the voice frequencies.

Sound effects, such as explosions or closing a door, contain a lot of low-frequency information. Most stereo systems do not respond to sub-audio frequencies, and therefore would benefit from a sub-woofer simulator.

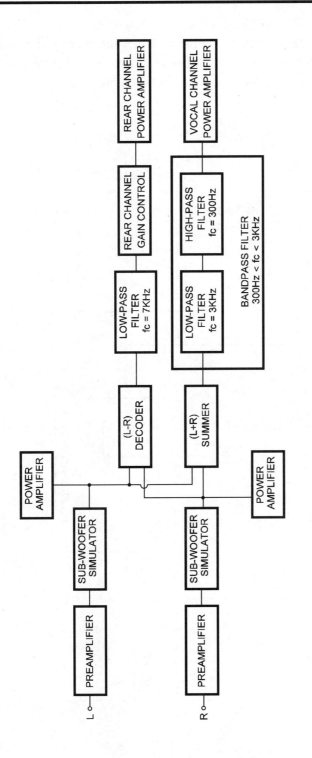

Figure 15-1.
Block diagram of the
home theater system.

The sub-woofer simulator boosts the low-frequency response of the system. The simulator consists of an input amplifier, a low-pass filter, and a summing amplifier as shown in *Figure 15-2*, which is a block diagram of the sub-woofer simulator.

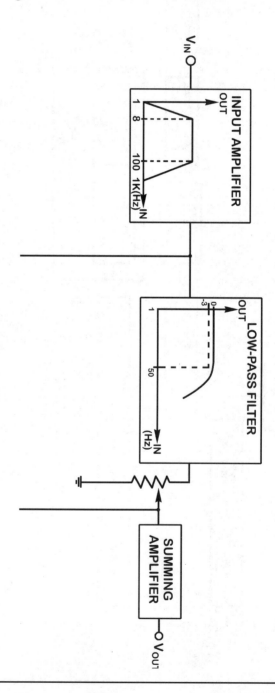

Figure 15-2.
Block diagram of the sub-woofer simulator.

CIRCUIT DESCRIPTION

The schematic of the home theater surround-sound system is shown in *Figures 15-3a* and *15-3b*. The left and right channel signal sources are selected by switch S1 and fed to integrated circuits U1A and U2A, which are configured as preamplifiers with a gain of:

$$A_v = 1 + (\frac{R1}{R2})$$

The gain is 11 with the components chosen. Integrated circuits U1A and U2A are configured as non-inverting amplifiers. The non-inverting amplifier has a high input impedance and a low output impedance. The preamplifier output signals may be varied by potentiometers RP1 and RP101. A dual potentiometer may be substituted for potentiometers RP1 and RP101. The LM324 quad operational amplifier was selected for U1, U2, U7 and U9 because the LM324 has four operational amplifiers.

Sound effects, such as explosions or thunder, contain a lot of information at frequencies below 50 Hz. Most amplifiers roll-off frequencies below 100 Hz; therefore, very low frequencies must be amplified, which requires a sub-woofer simulator to be incorporated into each front channel. This enables the amplifiers to reproduce the information present at low frequencies.

Operational amplifiers U1B, U1C and U1D form the left channel sub-woofer simulator, while operational amplifiers U2B, U2C and U2D form the right channel sub-woofer simulator. The sub-woofer simulator provides unity (0 dB) gain at frequencies above 1 kHz and it provides up to 28 dB of gain at 10 Hz.

Capacitors C1 and C101 are coupling capacitors and are used to block DC signals from entering the sub-woofer simulator. At very low frequencies, capacitors C1, C101, C2 and C102 act as open circuits. The gain of the input amplifiers are therefore less than unity at very low frequencies. At very high frequencies, capacitors C1, C101, C2 and C102 act as short circuits. The gain of the input

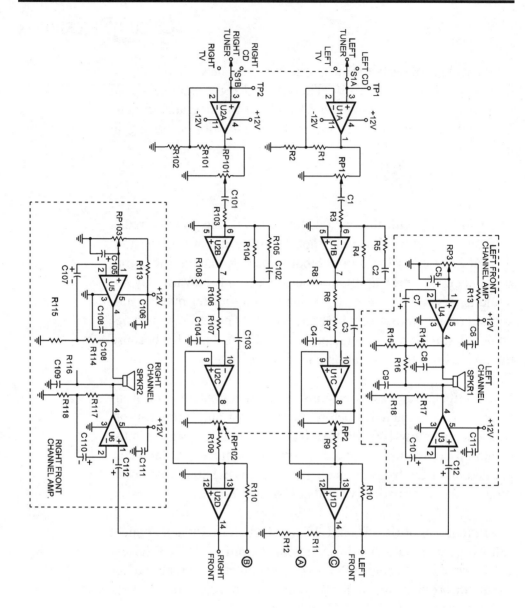

Figure 15-3a.
Schematic of the
home theater system.

amplifiers are therefore unity at higher frequencies. The bass boost is adjustable via potentiometers RP2 and RP102. A dual potentiometer may be substituted for potentiometers RP2 and RP102.

The decoder derives a rear-channel (L-R) signal from the two front channel signals. High frequencies are attenuated by a low-pass filter. The rear-channel amplitude is varied by the rear gain control RP4. The decoder passes the left input signal in phase and it inverts the right input signal.

The low-pass filter blocks high frequency signals. The low-pass filter is a second-order unity gain Sallen and Key filter. High frequency sounds are directional. In live performances, high frequency sounds are heard only when the listener is on-axis with the high frequency source.

Operational amplifier U7A is the decoder (L-R), a unity gain differential amplifier, where the left channel signal is fed to the non-inverting input of the operational amplifier and the right channel is fed into the inverting input of the operational amplifier. Operational amplifier U7B is configured as a 7000-Hz equal-component low-pass filter. The rear-channel gain can be adjusted with potentiometer RP4.

The rear-channel power amplifier can be any amplifier capable of driving two eight-ohm speakers in parallel. The LM383 (U8 of *Figure 15-3b*) is a cost-effective power amplifier suitable for audio applications. High current capability (3.5A) enables the device to drive low impedance loads with low distortion. The LM383 is current limited and thermally protected. A 5-pin TO-220 package is shown in *Figure 15-4*. The LM383 is pin-for-pin compatible with the TDA2002. The maximum power supply voltage allowed for the LM383 integrated circuit is 14 VDC. In *Figure 15-3b*, the LM383 is used as the rear channel power amplifier, U8, and as the voice channel power amplifier, U10.

The LM383 can deliver eight watts when used as a basic power amplifier. The rear channel and voice channel amplifiers are eight-

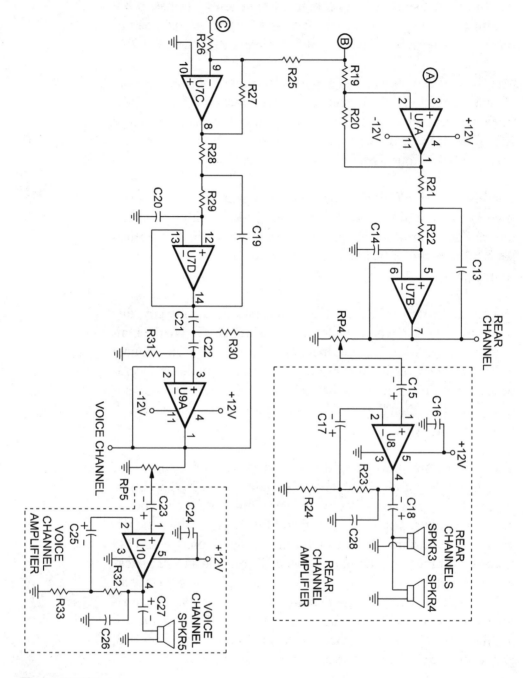

Figure 15-3b.
Schematic of the
home theater system
(continued).

watt power amplifiers. The LM383 can deliver 16 watts when used as a bridge amplifier. The left front and right front channel amplifiers are 16-watt power amplifiers. Two LM383 integrated circuits are required for a 16-watt bridge power amplifier. The left-channel bridge power amplifier consists of integrated circuits U3 and U4 and associated components. Potentiometer RP3 serves as the left-channel volume control. The right-channel bridge power amplifier consists of integrated circuits U5 and U6 and associated components. Potentiometer RP103 serves as the right-channel volume control. A dual potentiometer may be substituted for potentiometers RP3 and RP103.

The voice channel decoder consists of U7C, which is configured as a summing amplifier. It adds the left and right signals, resulting in a monophonic output signal. Only the voice frequencies (300 to 3000 Hz) are passed through the bandpass filter, which consists of integrated circuits U7D and U9A. The fourth-order Sallen and Key bandpass filter consists of a second-order unity-gain Sallen and Key low-pass filter, U7D, connected in series with a second-order unity-gain Sallen and Key high-pass filter, U9A. The voice-channel power amplifier consists of U10 and associated components. Potentiometer RP5 is the voice-channel volume control.

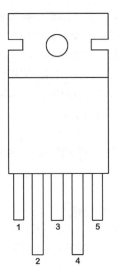

Figure 15-4.
Pin configuration
of the LM 383
integrated circuit.

PIN 1: NON-INVERTING INPUT
PIN 2: INVERTING INPUT
PIN 3: GROUND
PIN 4: OUTPUT
PIN 5: POSITIVE POWER SUPPLY

The bipolar power supply for the home theater system is shown in *Figure 15-5*. Transformer T1 steps down the household alternating line voltage to a low alternating voltage. Diodes D1-D4 rectify the low alternating voltage into a pulsating DC voltage. Capacitors C29-C34 smooth the pulsating DC voltage to a DC voltage with a very low AC or ripple component.

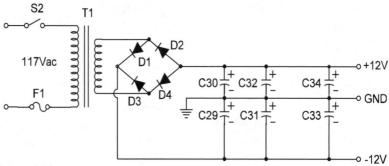

Figure 15-5.
Power supply schematic.

CONSTRUCTION

The surround-sound decoder may be built on a piece of perf board using point-to-point wiring. *Figure 15-6* is a diagram of the pin configuration of the LM324 quad operational amplifier integrated circuit used. Each LM324 contains four separate operational amplifiers. Alternately, thirteen LM741 operational amplifiers may be used as long as the proper pins are used for the inverting input, non-inverting input, output and the two power supply leads.

The surround-sound decoder is a moderately complex project, as can be seen from the parts list in *Table 15-1*.

Care should be taken to orient the capacitors, diodes and integrated circuits properly and to avoid cold solder joints and short circuits created by solder bridges. Most problems are caused by cold solder joints and solder bridges.

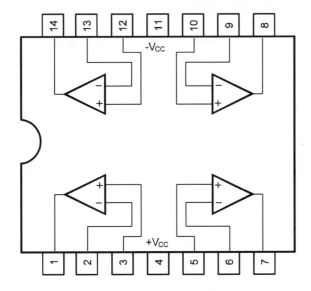

Figure 15-6.
Pin diagram of the
LM324 integrated circuit.

TESTING AND ADJUSTING

Testing of the decoder is straightforward. Connect the same signal to TP#1 and TP#2. There should be a signal at the outputs of U1D and U2D. There should be no signal at the output of U7A and no sound from the rear speakers. There should be a signal at the output of U7C and sound from the voice-channel speaker. If either TP#1 or TP#2 is grounded and a signal is fed to the other input, there should be a signal present at the outputs of U7A and U7C. There should be sound from the rear speakers and from the voice-channel speaker. These initial tests should be conducted using a 1-kHz sinewave for the test signal.

Signals lower than 100 Hz will be amplified and can be adjusted using bass controls RP2 and RP102. The gain of the rear channel can be adjusted with gain control RP4. The frequency response of the rear channel should roll-off above 7000 Hz.

All resistors are 1/4 watt at 5% unless otherwise noted.

R1, R101, R8, R108, R10, R110, R11, R12, R19, R20, R25-R27: 10K

R2, R102: 1K

R3, R103: 47K

R4, R104: 270K

R5, R105: 56K

R6, R106, R7, R107: 33K

R9, R109: 1.8K

R13, R113: 1.0M

R14, R114, R16, R116, R17, R117, R23, R32: 220 ohms

R15, R115, R18, R118, R24, R33: 10 ohms

R21, R22: 22K

R28-R31: 51K

RP1, RP101, RP4, RP5: 5K audio taper potentiometer

RP2, RP102, RP3, RP103: 100K audio taper potentiometer

C1, C101: 1.0 uF at 25V

C2, C102: 0.047 uF

C3, C4, C103, C104: 0.1 uF

C5, C105, C12, C112, C15, C23: 10 uF at 25V

C6, C106, C8, C108, C9, C109, C11, C111, C16, C24, C26, C28: 0.22 uF

C7, C107, C10, C110, C17, C25: 470 uF at 25V

C13, C14, C19, C20: 0.001 uF

C18, C27: 2000 uF at 25V

C21, C22: 0.01 uF

C29-C34: 1000 uF at 25V

D1-D4: MR821 5A rectifier diode

T1: **18** V.C.T. at 5A secondary step-down power transformer

U1, U2, U7, U9: LM324 quad operational amplifier

U3-U6, U8, U10: LM383 or TDA2002 eight-watt audio power amplifer

S1: 2P3T rotary switch

S2: SPST switch

F1: 5A slow-blow fuse

SPKR1, SPKR2, SPKR5: 4- to 8-ohm speaker systems

SPKR3, SPKR4: 8-ohm speaker systems

Table 15-1. Parts list for the home theater system.

Modifications

The left front and right front channel power amplifiers and associated components (shown in dashed boxes in *Figure 15-3a*) can be deleted and replaced with a stereo amplifier. The rear and voice channel power amplifiers and associated components (shown in dashed boxes in *Figure 15-3b*) can be deleted and replaced with a second stereo amplifier. If the second stereo power amplifier has volume controls, the rear channel gain control, RP4, and the voice channel gain control, RP5, may be deleted.

If the four power amplifiers shown in Figures 15-3a and *15-3b* are replaced with two stereo amplifiers, or if the four power amplifiers are built on a separate chassis with a separate single-ended 12-volt DC (at 5 amperes) power supply, the current rating of the secondary of transformer T1 can be reduced to one ampere. Diodes D1-D4 can be replaced with four 1N4004 rectifier diodes and fuse F1 can be replaced with a one ampere slow-blow fuse.

Installation

The placement of the rear speakers is important. They should be mounted at the rear of the side walls about six feet above the floor. This is shown in *Figure 15-7*. If the speakers are mounted on the rear wall, they will not sound like ambient reflected sound. Proper placement should bring exciting three-dimensional surround sound into your listening room.

Figure 15-7.
Proper rear
speaker placement.

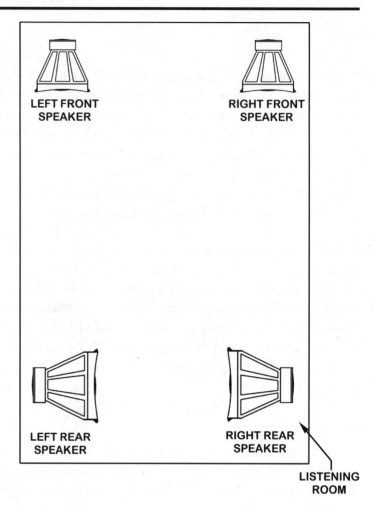

Chapter 16

Ultrasonic
Remote Control

It is often necessary to switch a device ON or OFF from a distance. This is especially useful if the device is remotely located, such as a satellite in space. For these applications, a remote control system is required. A remote control system consists of a transmitter and a receiver. It is often necessary, for security reasons, to encode the signal at the transmitter and then to decode the signal at the receiver. The transmitter generates a signal which is then sent into space. The receiver, which in this case is in the satellite, receives the signal and converts it into an appropriate command, such as turning a device ON or OFF.

Radio frequency (RF) remote control, *infrared* (IR) remote control and *ultrasonic* remote control are three types of remote control systems. Radio frequency systems are the first choice for remote locations, such as outer space. Infrared systems are the choice for line-of-sight applications and ultrasonic systems are the choice for locations that are not line-of-sight. Ultrasonic systems are also useful for line-of-sight applications.

This ultrasonic remote control has a range of about 15 feet. The signal is encoded by the transmitter and then the signal is decoded by the receiver. The transmitter circuit is compact enough to fit into a small pocket-size container.

TRANSMITTER CIRCUIT DESCRIPTION

The schematic of the ultrasonic transmitter is shown in *Figure 16-1*. The transmitter is a two-transistor oscillator that oscillates at 41.5 kHz. Transistor Q1 is biased by resistors R1 and R2 in a collector-to-base biasing circuit. Capacitor C1 is a speed-up capacitor—it reduces the turn-on and turn-off times of Q1 because it bypasses R1. Capacitor C2 is a coupling capacitor. Transistor Q2 is biased by resistors R3-R5 in a self-bias circuit. Capacitor C3 provides a shunt-series positive feedback path from the output of Q2 to the input of Q1. The output of Q2 is in phase with the input of Q1. The gain of the two-stage transistor amplifier is increased by C3, resulting in a two-stage transistor oscillator. Capacitor C4 is a shunt capacitor across the output of Q2. Capacitor C4 charges through resistor R5 when Q2 is in its cutoff mode. Capacitor C4 discharges through transistor Q2 when Q2 is in its saturation mode. Capacitor C5 is a coupling capacitor. The output of Q2 drives the ultrasonic transducer UT1, which converts the oscillator output signal into ultrasonic sound waves. Capacitor C6 is a filter capacitor.

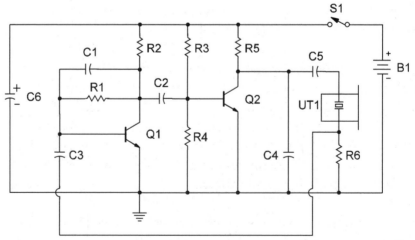

Figure 16-1.
Schematic of the
ultrasonic transmitter.

RECEIVER CIRCUIT DESCRIPTION

Figure 16-2 is the block diagram of the ultrasonic receiver. It consists of an ultrasonic transducer, a high-gain two-transistor amplifier, a rectifier, a comparator, a decoder and latch, and a transistor switch which drives the load.

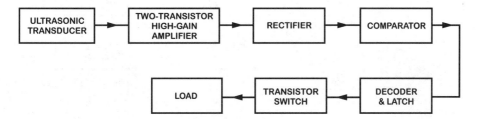

Figure 16-2.
Block diagram of
the ultrasonic receiver.

The schematic of the receiver is shown in *Figure 16-3*. The ultrasonic transducer UT1 converts the ultrasonic sound waves into an electrical signal. The two-transistor amplifier provides a gain of about 70 decibels. Resistors R2, R4 and R5 bias transistors Q1 and Q2 in self-bias circuits. Resistors R1 and R3 provide a DC negative shunt-series feedback path for increased amplifier stability. Capacitor C1 is a decoupling capacitor, which provides an AC negative feedback path for increased amplifier stability.

Diode D1 rectifies the output signal of transistor Q2. The diode passes only the negative half-cycles of the output signal of Q2. The rectified signal is applied to the inverting input of integrated circuit U1, which is configured as a comparator. The LM301 has a high slew rate and it is not compensated. Compensating an operational amplifier decreases its slew rate. A high slew rate is essential for comparator operation. Resistor R9 provides a shunt-shunt positive feedback path around U1 to avoid false triggering of the relay. The reference voltage at the non-inverting input of the comparator can be adjusted by potentiometer RP1.

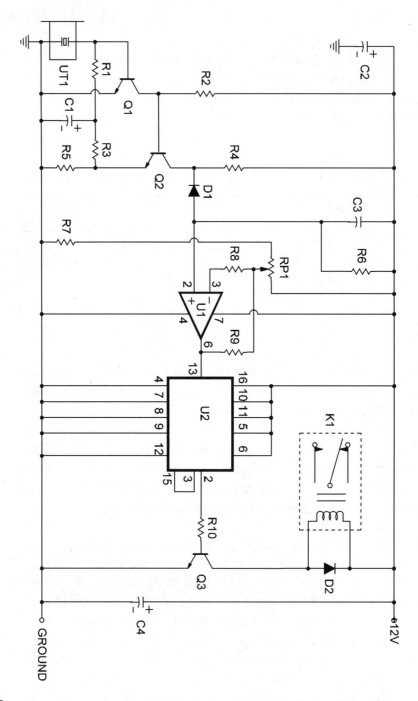

Figure 16-3.
Schematic of
the ultrasonic receiver.

If the signal applied to the inverting input of U1 increases, the output voltage of the comparator becomes negative. If the signal applied to the inverting input of U1 becomes very large, comparator U1 cuts off the output signal. If the voltage at the inverting input of U1 is less than the reference voltage at the non-inverting input of U1, the output of comparator U1 goes HIGH (about 10.5V) and serves as a positive-going or leading-edge clock pulse to integrated circuit U2, which is configured as a decoder and latch. U2 is pulsed only on the leading edge of the clock pulse. If the voltage at the inverting input of U1 is greater than the reference voltage at the non-inverting input of U1, the output of comparator U1 goes LOW and integrated circuit U2 is not pulsed because the clock pulse is a negative-going or trailing-edge clock pulse.

Integrated circuit U2 serves as a latch. The output (Q) of the first flip-flop changes state (from HIGH to LOW) on the leading edge of the first clock pulse. The second flip-flop is not clocked because Q is now a trailing-edge clock pulse. On the leading edge of the second clock pulse, the first flip-flop changes state (from LOW to HIGH) and clocks the second flip-flop because Q is now a leading-edge clock pulse. The inverted output (Q´) of the second flip-flop changes state (LOW to HIGH) and drives transistor switch Q3.

Integrated circuit U2 also serves as a decoder. The output of the first flip-flop (Q) serves as the clock pulse for the second flip-flop. The receiver must detect two signals for each command signal that it generates. The inverted output of the second flip-flop of U2 supplies a base drive current to transistor switch Q3 through current limiting resistor R10. Transistor Q3 drives relay K1. Diode D2 protects transistor Q3 from any inductive spike voltages generated when the relay coil is powered. Relay K1 is used to switch a device ON or OFF.

The schematic of the power supply for the ultrasonic receiver is shown in *Figure 16-4*. Transformer T1 steps down the household alternating line voltage to a low alternating voltage. Diodes D1 and D2 rectify the low alternating voltage into a pulsating DC voltage. Capacitor C1 smooths the pulsating DC voltage into a DC

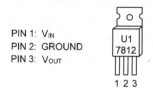

PIN 1: V_{IN}
PIN 2: GROUND
PIN 3: V_{OUT}

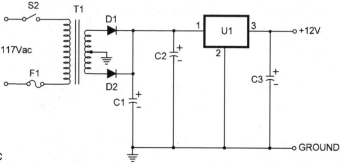

Figure 16-4.
Schematic of the ultrasonic
receiver power supply.

voltage with a small ripple or AC component. Capacitor C2 is required if the positive voltage regulator, U1, is located far from filter capacitor C1. Voltage regulator U1 reduces the AC component of the DC voltage. Capacitor C3 improves the transient response of U1.

CONSTRUCTION

The transmitter and receiver circuits are built on separate pieces of perf board. All diodes, capacitors, transistors and integrated circuits must be properly oriented. Integrated circuit sockets should be used for the two integrated circuits in the receiver. The parts list for the ultrasonic transmitter is given in *Table 16-1*. The parts list for the ultrasonic receiver is given in *Table 16-2*, and the parts list for the power supply is given in *Table 16-3*.

Care should be taken when soldering the components. Most problems are caused by cold solder joints or by short circuits created by solder bridges. Integrated circuit sockets should be used for integrated circuits U1 and U2 of the receiver.

The pin configuration and truth table for the CMOS CD4027 dual J-K master-slave flip-flop IC is shown in *Figure 16-5*. The pin configuration of the LM301 operational amplifier is shown in *Figure 16-6*.

ADJUSTMENT

The sensitivity of the ultrasonic receiver is adjusted with potentiometer RP1. Connect a DC voltmeter across diode D2. Adjust potentiometer RP1 until the DC voltmeter reads the same voltage as the receiver power supply voltage. The relay should "click" ON.

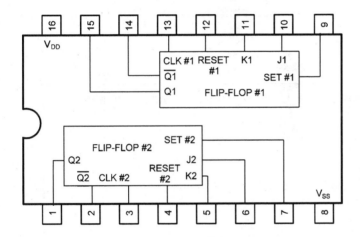

Figure 16-5.
Pin configuration
of the CD4027
dual J-K flip-flop.

PRESENT STATE				OUTPUT	CLOCK	NEXT STATE	
INPUTS				OUTPUT	CLOCK	OUTPUTS	
J	K	S	R	Q		Q	Q̄
1	X	0	0	0	⌐⌐	1	0
X	0	0	0	1	⌐⌐	1	0
0	X	0	0	0	⌐⌐	0	1
X	1	0	0	1	⌐⌐	0	1
X	X	0	0	X	⌐⌐	NO CHANGE	NO CHANGE
X	X	1	0	X	X	1	0
X	X	0	1	X	X	0	1
X	X	1	1	X	X	1	1

TRUTH TABLE

NOTES:

① X = DON'T CARE STATE

② Q' = Q̄

Slowly adjust RP1 until the DC voltmeter reads zero volts. The relay should "click" OFF. The receiver is now adjusted for maximum sensitivity. The range for the ultrasonic remote control is about 15 feet. The ultrasonic remote control should provide years of trouble-free service.

Figure 16-6.
LM301 operational amplifier pin configuration.

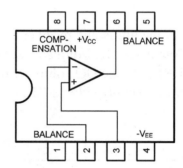

All resistors are 1/4W at 5% unless otherwise noted.
R1: 120K
R2: 6.8K
R3: 100K
R4: 8.2K
R5: 1K
R6: 2.7 ohms
C1: 0.001 uF
C2: 0.01 uF
C3: 0.047 uF
C4: 680 pF
C5: 0.0033 uF
C6: 100 uF at 25V
Q1, Q2: 2N3904
UT1: ultrasonic transmitter transducer
B1: 9-volt battery
S1: SPST switch

Table 16-1. Parts list for the ultrasonic transmitter.

All resistors are 1/4W at 5% unless otherwise noted.

R1, R4: 10K
R2, R6, R8: 100K
R3: 100 ohms
R5: 2.2K
R7: 68K
R9: 1M
R10: 4.7K
RP1: 100K trimmer potentiometer
C1: 1.0 uF at 25V
C2: 10 uF at 25V
C3: 0.047 uF
C4: 470 uF at 25V
Q1, Q2, Q3: 2N3904
U1: LM301 operational amplifier
U2: CD4027 CMOS dual J-K master-slave flip-flop
D1: 1N914 switching diode
D2: 1N4001 rectifier diode
K1: 12V relay
UT1: ultrasonic receiver transducer

Table 16-2. Parts list for the ultrasonic receiver.

C1: 1000 uF at 25V
C2: 0.33 uF tantalum
C3: 1.0 uF tantalum
D1, D2: 1N4001 rectifier diode
U1: 7812 positive voltage regulator
T1: 25.2 V.C.T. at 1A secondary step-down power transformer
S1: SPST switch
F1: 1A slow-blow fuse

Table 16-3. Parts list for the ultrasonic receiver power supply.

Chapter 17

Ultrasonic Radar for Automobiles

Many people park their cars in small spaces "by feel." They back up and/or drive forward until they gently bump the car in front or behind. This ultrasonic radar eliminates parking by feel.

The transmitter continuously emits ultrasonic sound waves. When your car is within 12 inches of the next car, the receiver detects the reflected ultrasonic sound waves and a dash-mounted light-emitting diode (LED) illuminates.

Radio frequency (RF) remote control, *infrared* (IR) remote control and *ultrasonic* remote control are three types of remote control systems. Radio frequency systems are the first choice for remote locations, such as outer space. Infrared systems are the choice for line-of-sight applications. Ultrasonic systems are the choice for locations that are not line-of-sight. Ultrasonic systems are also useful for line-of-sight applications.

CIRCUIT DESCRIPTION

Figure 17-1 is the block diagram of the ultrasonic radar for automobiles. It consists of a transmitter and a receiver. The receiver contains an ultrasonic receiver transducer, a two-transistor high-gain amplifier, a rectifier, comparator, transistor switch, and indicator.

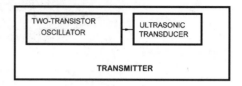

Figure 17-1.
Block diagram of
the ultrasonic radar.

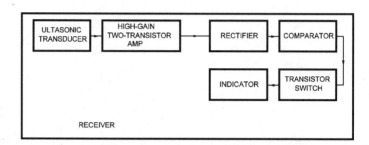

The schematic of the ultrasonic radar for automobiles is shown in *Figure 17-2*. The transmitter is a two-transistor oscillator that oscillates at 41.5 kHz. Transistor Q1 is biased by resistors R1 and R2 in a collector-to-base biasing circuit. Capacitor C1 is a speed-up capacitor—it reduces the turn-on and turn-off times of Q1 because it bypasses R1. Capacitor C2 is a coupling capacitor. Transistor Q2 is biased by resistors R3-R5 in a self-bias circuit. Capacitor C3 provides a shunt-series positive feedback path from the output of Q2 to the input of Q1. The output of Q2 is in phase with the input of Q1. The gain of the two-stage transistor amplifier is increased by C3, resulting in a two-stage transistor oscillator. Capacitor C4 is a shunt capacitor across the output of Q2. Capacitor C4 charges through resistor R5 when Q2 is in its cutoff mode. Capacitor C4 discharges through transistor Q2 when Q2 is in its saturation mode. Capacitor C5 is a coupling capacitor. The output of Q2 drives the ultrasonic transducer UT1, which converts the oscillator output signal into ultrasonic sound waves. Capacitor C6 is a filter capacitor.

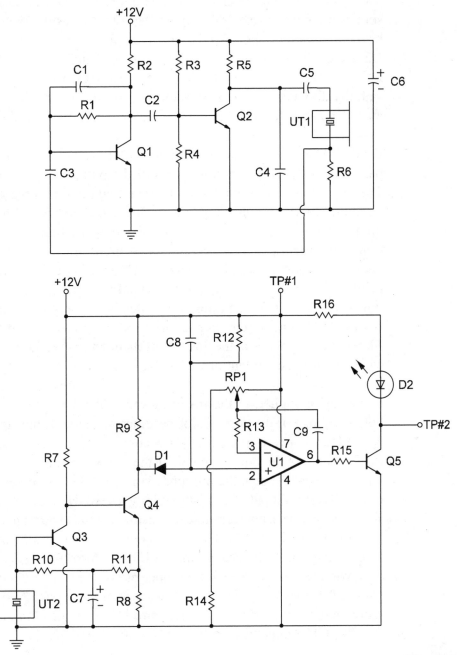

Figure 17-2.
Schematic of the
ultrasonic radar.

The ultrasonic transducer UT2 converts the reflected ultrasonic sound waves into an electrical signal. The two-transistor amplifier provides a gain of about 70 dB. Resistors R7-R9 bias transistors Q3 and Q4 in self-bias circuits. Resistors R10 and R11 provide a DC negative shunt-series feedback path for increased amplifier stability. Capacitor C7 is a decoupling capacitor which provides an AC negative feedback path for increased amplifier stability.

Diode D1 rectifies the output signal of transistor Q4. The diode passes only the negative half-cycles of the output signal of Q4. The rectified signal is applied to the inverting input of integrated circuit U1, which is configured as a comparator. The LM301 has a high slew rate and it is not compensated. Compensating an operational amplifier decreases its slew rate. A high slew rate is essential for comparator operation. Capacitor C9 provides a shunt-shunt positive feedback path around U1 to avoid false triggering of the transistor switch Q5. The transistor switch is ON only when there is an output signal at the output (pin 6) of integrated circuit U1. The reference voltage at the non-inverting input of the comparator can be adjusted by potentiometer RP1.

If the signal applied to the inverting input of U1 increases, the output voltage of the comparator becomes negative. If the signal applied to the inverting input of U1 becomes very large, comparator U1 cuts off the output signal. If the voltage at the inverting input of U1 is less than the reference voltage at the non-inverting input of U1, the output of comparator U1 goes HIGH (about 10.5V) supplying a base drive current to transistor switch Q5 through current limiting resistor R15. The warning indicator, D2 illuminates. If the voltage at the inverting input of U1 is greater than the reference voltage at the non-inverting input of U1, the output of comparator U1 goes LOW. No base drive current is supplied to transistor switch Q5. The warning indicator D2 does not illuminate. Resistor R16 limits the current flow through LED D2.

A buzzer may be substituted for resistor R16 and LED D2. The positive lead of the buzzer is connected to TP#1. The negative lead of the buzzer is connected to TP#2.

CONSTRUCTION

The transmitter and receiver circuits are built on one piece of perf-board. All diodes, capacitors, transistors and the integrated circuit must be properly oriented. An integrated circuit socket should be used. The parts list for the ultrasonic transmitter is given in *Table 17-1*.

All resistors are 1/4W at 5% unless otherwise noted.
All capacitors are rated at 25 volts.

R1: 120K
R2: 6.8K
R3, R7, R12, R13: 100K
R4: 8.2K
R5,R16: 1K
R6: 2.7 ohms
R8: 2.2K
R9, R10: 10K
R11: 100 ohms
R14: 68K
R15: 4.7K
RP1: 100K trimmer potentiometer
C1: 0.001 uF
C2: 0.01 uF
C3,C8: 0.047 uF
C4: 680 pF
C5: 0.0033 uF
C6: 1000 uF
C7: 1.0 uF
C9: 10 uF nonpolarized capacitor
U1: LM301
Q1-Q5: 2N3904
D1: 1N914 switching diode
D2: light-emitting diode
UT1: ultrasonic transmitter transducer
UT2: ultrasonic receiver transducer

Table 17-1. Parts list for the ultrasonic radar.

Care should be taken when soldering the components. Most problems are caused by cold solder joints or by short circuits created by solder bridges.

The pin configuration of the LM301 operational amplifier is shown in *Figure 16-6*.

ADJUSTMENT AND TESTING

The sensitivity of the ultrasonic receiver is adjusted with potentiometer RP1. Connect the positive lead of a DC voltmeter to TP#1. Connect the negative lead of the DC voltmeter to TP#2. Adjust potentiometer RP1 until the DC voltmeter reads the same voltage as the radar power supply voltage. The LED should illuminate. Slowly adjust RP1 until the DC voltmeter reads zero volts. D2 should then extinguish. The receiver is now adjusted for maximum sensitivity.

Connect the ultrasonic radar to a 12-volt DC power supply. The ultrasonic transducers should be about 12 inches from a wall. The LED should illuminate. If the ultrasonic radar is moved away from the wall, the LED should extinguish. Adjust RP1 until the range for the ultrasonic radar for automobiles is about one foot.

INSTALLATION

The ultrasonic transducers can be mounted on the rear bumper, not more than six inches apart, as shown in *Figure 17-3*. Two holes should be drilled into the bumper and the ultrasonic transducers can then be glued with epoxy into place. The electronics of the ultrasonic radar should be installed in a suitable case and mounted in the trunk, as close as possible to the ultrasonic transducers. Coaxial cables should be used to connect the ultrasonic transducers to the electronics of the radar. The LED warning indicator can be mounted on the dashboard. Twin-lead speaker wire may be used to connect the warning indicator to the electronics of the radar. The positive power lead of the radar is connected to positive side

1. **Warning indicator (dash mounted)**
2. **Ultrasonic transmitter transducer**
3. **Ultrasonic receiver transducer**
4. **Electronics of the ultrasonic radar**

Figure 17-3.
Installing the ultrasonic
radar in an automobile.

of the car's battery via the ignition switch. The negative lead of
the radar is connected to the electrical ground of the automobile.

If the rear bumper of the automobile is hollow, three holes should
be drilled in the bumper, as shown in *Figure 17-4*. The electronics
of the radar and the transducers can be built into a plastic box and
mounted inside the hollow rear bumper. The radar assembly is
fastened to the bumper with a mounting bolt, lockwasher and nut.
The power leads are connected to the 12-volt power supply of the
automobile. The LED is mounted on the dashboard and connected
to the radar with twin-lead speaker wires.

A second ultrasonic radar may be built to monitor the front bumper
of the car. The ultrasonic radar should provide years of trouble-
free service.

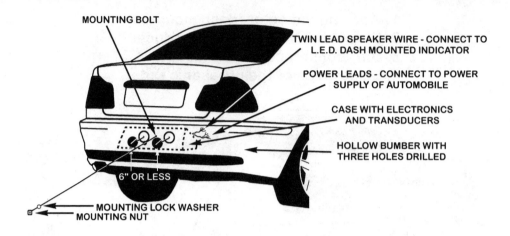

MOUNTING BOLT

TWIN LEAD SPEAKER WIRE - CONNECT TO
L.E.D. DASH MOUNTED INDICATOR

POWER LEADS - CONNECT TO POWER
SUPPLY OF AUTOMOBILE

CASE WITH ELECTRONICS
AND TRANSDUCERS

HOLLOW BUMBER WITH
THREE HOLES DRILLED

6" OR LESS

MOUNTING LOCK WASHER
MOUNTING NUT

Figure 17-4.
Installing the radar
inside a hollow rear bumper.

Chapter 18

Dual Audio Preamplifier Integrated Circuits

The LM38x is a versatile series of linear dual preamplifier integrated circuits. They are well suited for audio applications such as preamplifiers and tone control circuits. The LM38x series of integrated circuits has excellent power supply ripple rejection, low distortion, wide bandwidth and low noise. The LM38x series requires a single-ended power supply.

The LM381 allows external noise figure optimization (for narrow-band applications) and compensation (for low-gain applications). The LM381 is usually used as a differential input amplifier. It can also be used as a single-ended input amplifier in ultra-low-noise circuits. The LM381A is a low-noise version of the LM381.

The LM382 cannot be externally compensated and it cannot be used as a single-ended input amplifier. The LM382 has an internal resistor matrix which allows the user to select from several closed-loop gain and frequency-response options.

The LM387 has only input, output and power supply terminals. The LM387 cannot be externally compensated and it cannot be used as a single-ended input amplifier. The LM387A is a low-noise version of the LM387.

LM381/LM381A

The LM381 is a dual preamplifier designed to amplify low-level signals in applications requiring optimum noise performance. Each of the preamplifiers is independent, with its own internal power supply decoupler and regulator, providing 120 decibels of power supply ripple rejection. There is 60 decibels of channel separation between the two preamplifiers. Each preamplifier has 112 decibels of gain, a wide unity-gain bandwidth of 15 MHz and a large output voltage swing of (Vcc-2) volts peak-to-peak.

The schematic of one of the preamplifiers and the pin configuration of the LM381 are shown in *Figure 18-1*. The differential input amplifier consists of transistors Q1, Q2 and associated components.

The second stage consists of Q3-Q6. Transistors Q3 and Q4 form a darlington-emitter follower which has a voltage gain of 2000. Capacitor C1 internally compensates the preamplifier for a gain of unity at 15 MHz, giving stability for closed-loop gains of 10 or more. The output of the darlington-emitter follower drives the level shifter, which consists of transistors Q5 and Q6. Transistor Q5 is configured as a common-emitter, and Q6 is a constant-current load.

The level shifter drives the output stage, which consists of Q7-Q10. The output stage consists of a darlington-emitter follower Q8 and Q9 and an active current sink Q7. Transistor Q10 limits the output current to 12 milliamperes.

The bias network consists of transistors Q11-Q15. The bias network yields 120 decibels of power supply ripple rejection. Transistors Q11-Q13 form a high impedance current generator, which generates a DC reference voltage across diode D3. The reference voltage operates the first two stages via Q14 and Q15 and biases the base of Q1 internally.

The LM381 leads are connected to an external circuit depending on the circuit application.

Pin #1: non-inverting input to preamplifier #1

Pin #2: inverting input when preamplifier #1 is used as a differential input amplifier

Pin #3: inverting input when preamplifier #1 is used as a single-ended input amplifier

Pin #4: connect to ground

Pin #5: external compensation for preamplifier #1

Pin #6: external compensation for preamplifier #1

Pin #7: output of preamplifier #1

Pin #8: output of preamplifier #2

Pin #9: connect to the positive rail of a 9V-40V DC power supply

Pin #10: external compensation for preamplifier #2

Pin #11: external compensation for preamplifier #2

Pin #12: inverting input when preamplifier #2 is used as a single-ended input amplifier

Pin #13: inverting input when preamplifier #2 is used as a differential input amplifier

Pin #14: non-inverting input to preamplifier #2

The LM381 can be used as a differential input amplifier by using resistors R1 and R2 in a DC negative-feedback loop as shown in *Figure 18-2*. The inverting input is biased externally at 1.2 volts by resistors R1 and R2. The non-inverting input goes to 1.2 volts. The output quiescent point is established by a DC negative-feedback loop through resistors R1 and R2. For bias stability, the current flow through R2 is 10 times the input current of Q2 (of the LM381 preamplifier), which is about 0.5 microamperes.

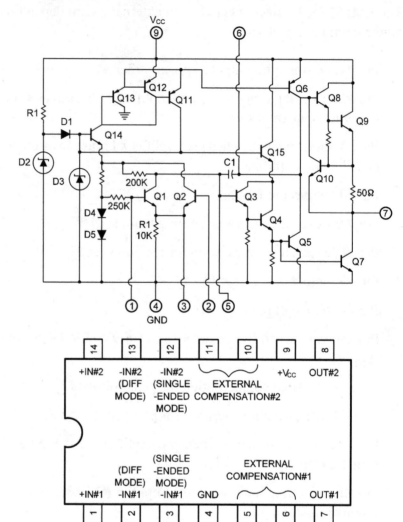

Figure 18-1.
Schematic and
pin configuration
of the LM381.

For a differential input:

$$R2 = \frac{2V_{BE}}{10I_{Q2}} = \frac{1.2}{5 \times 10^{-6}} = 240k \text{ ohms maximum}$$

and

$$R1 = [(\frac{Vcc}{2.4}) - 1]R2 \text{ ohms}$$

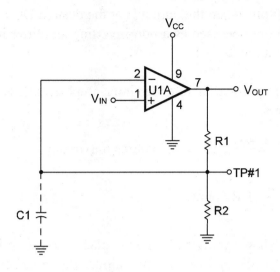

Figure 18-2.
Differential biasing
of the LM 381.

NOTES:

① U1A: LM381 (PREAMPLIFIER #1)

② TP#1 = 1.2V

③ $V_{OUT} = 1.2 (1 + \frac{R1}{R2})$

The DC voltage gain is:

$$A_v = \frac{R1}{R2}$$

and the AC voltage gain is:

$$A_v = 1 + (\frac{R1}{R2})$$

The LM381 can be used as a single-ended input amplifier by grounding the inverting differential input (pin #2 for preamp #1). The differential input stage is powered by an internal 5.6V regulator. The collector of Q1 is fed to the output via a DC amplifier. Transistor Q2 is disabled by grounding its base. The base of Q2 must still be biased using emitter feedback, shown in *Figure 18-2*. Capacitor C1 (shown in dashed lines) is used to ground the base of Q2. The inverting single-ended input (pin #3 for preamp #1) must be biased to 0.6 volts when the output is at the desired DC level. The LM381 can only be used as a non-inverting amplifier in its single-ended input mode.

The feedback current for a single-ended input amplifier is less than 100 microamperes:

$$R2 = \frac{V_{BE}}{I_{FB}} = \frac{0.6}{5} \times 10^{-4} = 1.2 \text{ kohms maximum}$$

$$R1 = (\frac{Vcc}{1.2} - 1)R2 \text{ ohms}$$

If capacitor C1 (shown in dashed lines) is placed across resistor R2 in *Figure 18-2*, the AC gain of the amplifier approaches the open-loop gain of the amplifier. The low-frequency cutoff frequency, f_o is:

$$f_o = \frac{A_o}{(2 \times PI \times R1C1)}$$

where: A_o is the open-loop gain of the amplifier.

LM382

The LM382 is identical to the LM381 except that the LM382 has an internal five-resistor matrix. The LM382 cannot be used in single-ended input applications and it cannot be externally compensated. The LM382 has a gain of 100 decibels. The resistor matrix simplifies bias and filter-network design. The resistor matrix is intended for use in circuits where the LM382 is powered by a 12-volt power supply.

The pin configuration of the LM382 is shown in *Figure 18-3*. The leads of the LM382 are connected to the external circuit depending on the circuit application.

Pin #1: non-inverting input to preamplifier #1

Pin #2: inverting input to preamplifier #1

Pins #3, 5 and 6: gain control for preamplifier #1

Pin #4: connect to ground

Pin #7: output of preamplifier #1

Pin # 8: output of preamplifier #2

Pins #9, 10 and 12: gain control for preamplifier #2

Pin #11: connect to the positive rail of a 9V-40V DC power supply

Pin #13: inverting input to preamplifier #2

Pin #14: non-inverting input to preamplifier #2

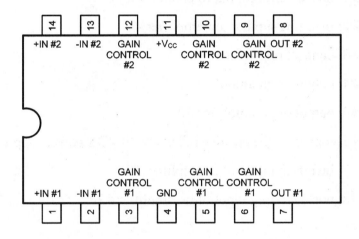

Figure 18-3.
Pin configuration
of the LM382.

LM387/LM387A

The LM387 is identical to an LM381 except that the LM387 cannot be externally compensated. The LM387 may be used as a differential input amplifier without external compensation. The LM387A is a low-noise version of the LM387. The pin configuration of the LM387 is shown in *Figure 18-4*.

The leads of the LM387 are connected to the external circuit depending on the circuit application.

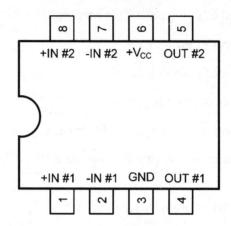

Figure 18-4.
Pin configuration
of the LM387.

Pin #1: non-inverting input to preamplifier #1

Pin #2: inverting input to preamplifier #1

Pin #3: connect to ground

Pin #4: output of preamplifier #1

Pin #5: output of preamplifier #2

Pin #6: connect to the positive rail of a 9-40 VDC power supply

Pin #7: inverting input of preamplifier #2

Pin #8: non-inverting input of preamplifier #2

APPLICATIONS

The LM38x series of linear dual preamplifier integrated circuits are high-gain wide-bandwidth devices. They are therefore subject to RF (radio frequency) instability and RF pickup.

RF instability can be eliminated by decoupling the power supply at high frequencies, by placing C1, which is a 0.1-uF ceramic capacitor or 0.001 uF tantalum capacitor, across the integrated circuit power supply and ground pins as shown in *Figure 18-5*.

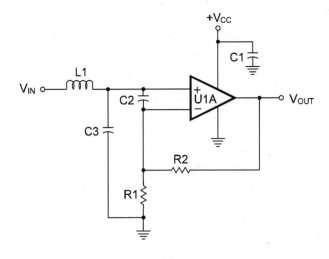

Figure 18-5.
Eliminating RF
instability and RF pickup.

L1: 10μH RF CHOKE
C1: 0.1μF CERAMIC OR 0.001μF TANTALUM
C2: 10pF TO 300pF CERAMIC
C3: 10pF TO 300pF CERAMIC
U1: LM381 OR LM382 OR LM387

RF pickup is AM modulation at the input of the preamplifier. RF pickup can be eliminated by placing L1, which is a 10-uH RF choke, in series with the preamplifier's input terminal, also shown in *Figure 18-5*. The input can also be decoupled by placing C2 and/or C3 which are 10 to 300-pF ceramic capacitors, as shown in *Figure 18-5*. It may not be necessary to use all three components (L1, C2 and C3) to eliminate the RF pickup.

An inverting amplifier is shown in *Figure 18-6*. The non-inverting input terminal is grounded by capacitor C2. The AC voltage gain is:

$$A_v = \frac{-R3}{R2} = -10$$

The quiescent output is 12 volts and the input impedance approximately equals R1.

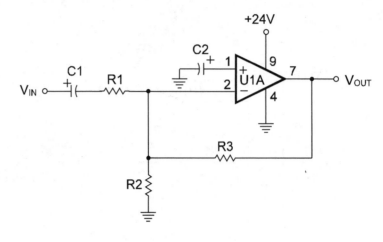

Figure 18-6.
LM381 inverting
amplifier.

$$A_V = \frac{-R3}{R1} = -10$$

C1, C2: 1μF
R1: 10K
R2: 12K
R3: 100K
U1: LM381

A non-inverting amplifier is shown in *Figure 18-7*. The input impedance is 250k ohms. To avoid distortion, the input signal must not exceed 300 millivolts RMS. The inverting input is floating because it is connected through a high impedance (R3 and C2) to ground. The DC voltage gain is:

$$\frac{R1}{R2} = 8.3$$

and the AC voltage gain is:

$$\frac{R1 + R3}{R3} = 1 + \frac{R1}{R3} = 101$$

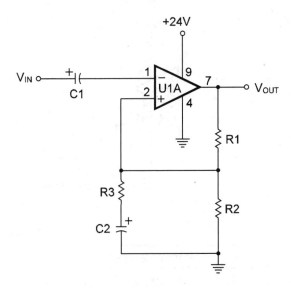

+24V

V$_{IN}$ o

C1

1
2 U1A
9
7 o V$_{OUT}$
4

R1

R3

R2

C2

Figure 18-7.
LM381 non-inverting
amplifier.

$$A_V = \frac{(R1 + R3)}{R3} = 101$$

C1, C2: 10 μF
R1: 1M
R2: 120K
R3: 10K
U1: LM381

An LM382 fixed-gain non-inverting amplifier is shown in *Figure 18-8*. The gain can be changed with capacitor(s) C1 and/or C2 as specified in *Figure 18-8*.

An LM382 inverting amplifier is shown in *Figure 18-9*. The non-inverting input is grounded by capacitor C2. The voltage gain of the amplifier is:

$$A_v = \frac{-2.7 \times 10^5}{R1} = -100$$

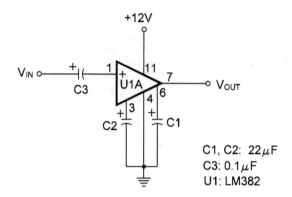

Figure 18-8.
LM382 fixed-gain
non-inverting amplifier.

GAIN	CAPACITOR USED
40dB	C1 ONLY
55dB	C2 ONLY
80dB	C1 AND C2

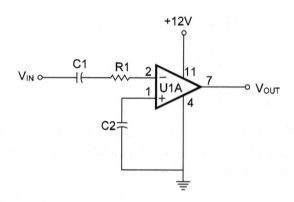

Figure 18-9.
LM382 inverting
amplifier.

$$A_V = \frac{-270{,}000}{R1} = -100$$

R1: 2.7K
C1, C2: 0.1μF
U1: LM382

An LM387 "speech" bandpass filter is shown in *Figure 18-10*. The bandpass filter consists of a second-order low-pass filter in series with a second-order high-pass filter. The frequency selective components were selected to give a 12 dB/octave attenuation to signals whose frequencies are outside the speech frequency range of 300 to 3000 Hz.

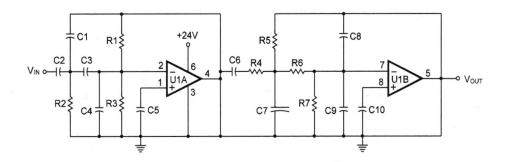

R1: 2.2M
R2: 430K
R3: 240K
R4, R5: 270K
R6: 130K
R7: 47K
C1, C7, C8, C2, C3: 560pF
C4: 0.01 μF
C5, C6, C10: 0.1 μF
C9: 0.0022 μF
U1: LM387

Figure 18-10.
LM387 "speech"
bandpass filter.

Chapter 19

Tone Control Circuits

The human ear is most sensitive to "speech" frequencies, especially at low sound intensities. Tone control circuits are required in audio equipment to boost frequencies below and above the "speech" frequencies of 300 Hz to 3000 Hz.

Baxandall bass and treble control circuits are very popular tone controls. Baxandall tone circuits are four-terminal networks. There is an input terminal, an output terminal, a ground terminal and a feedback-loop terminal. Baxandall tone controls are used in active tone control circuits.

Passive tone control circuits are three-terminal networks. There is an input terminal, an output terminal and a ground terminal.

Passive tone control circuits are usually placed between two preamplifier stages. The voltage gain of the preamplifiers should equal the insertion loss (available "boost") of the tone control circuit.

PASSIVE TONE CONTROL CIRCUITS

Passive networks have unity gain because they have no amplifiers. Resistors, capacitors and inductors are the only components used in passive networks.

A passive bass control circuit and its frequency response curves are shown in *Figure 19-1*.

The ratio of resistors R2/R1 and R1/RP1 determine the amount of bass "boost" and "cut." Since R2/R1 = 1/10 and R1/RP1 = 1/10, there is 20 dB of "boost" and -20 dB of "cut." The low-frequency control point, f1, is set where X_{C1} = R1 and X_{C2} = R2. With the components selected, f1 = 1600 Hz.

Capacitors C1 and C2 can be calculated:

$$C1 = \frac{1}{(2 \times PI \times f1 \times R1)} \quad \text{and} \quad C2 = \frac{1}{(2 \times PI \times f1 \times R2)}$$

The gain of the bass control is:

$$A = \frac{R2 + (RP1B // X^{C2})}{R1 + R2 + (RP1A // X^{C1}) + (RP1B // X^{C2})}$$

where: // means "in parallel" and where $X_C = 1/(2 \times PI \times f \times C)$

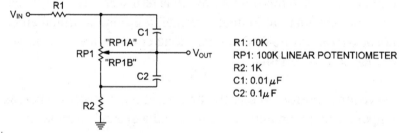

R1: 10K
RP1: 100K LINEAR POTENTIOMETER
R2: 1K
C1: 0.01 μF
C2: 0.1 μF

Figure 19-1.
Schematic and frequency response of bass control.

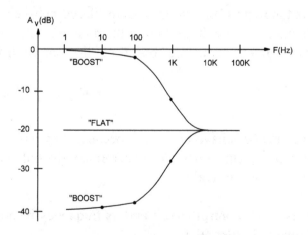

When the wiper of potentiometer RP1 is at the junction of RP1 and R2, the bass control gives the maximum bass "cut." When the wiper of potentiometer RP1 is at the junction of RP1 and R1, the bass control gives the maximum bass "boost." When the wiper of RP1 is at its mid-position, the bass control gives a "flat" response. The bass control is essentially a voltage divider.

A passive treble control circuit and its frequency response curves are shown in *Figure 19-2*. The ratio of capacitors C2/C1 determines the degree of treble "boost" and "cut." Since C2/C1 = 10, there is 20 dB of "boost" and -20 dB of "cut." The high-frequency control point, f2, is established where X_{C1} = RP1.

Capacitor C1 is calculated:

$$C1 = \frac{1}{(2 \times PI \times f2 \times RP1)}$$

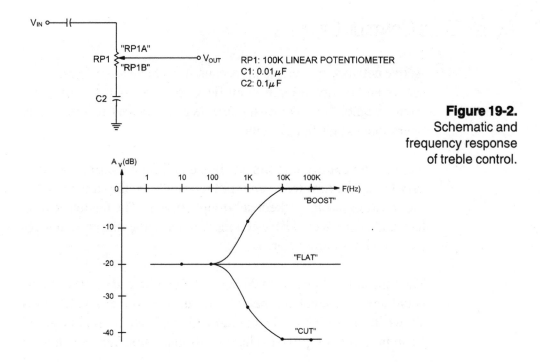

RP1: 100K LINEAR POTENTIOMETER
C1: 0.01 μF
C2: 0.1 μF

Figure 19-2.
Schematic and frequency response of treble control.

With the components selected, f2=1600 Hz. The gain of the treble control is;:

$$A_v = \frac{(RP1B + X_{C2})}{(RP1 + X_{C1} + X_{C2})}$$

where: RP1 = (RP1A + RP1B).

When the wiper of potentiometer RP1 is at the junction of RP1 and capacitor C2, the treble control gives the maximum treble "cut." When the wiper of RP1 is at the junction of RP1 and capacitor C1, the treble control gives the maximum treble "boost." When the wiper is in its mid-position, the treble control gives a "flat" response. The treble control is essentially a voltage divider.

The passive bass and treble control circuits can be combined as shown in *Figure 19-3*. The frequency response curves of the bass and treble control circuit are also shown in *Figure 19-3*. There is 20 dB of bass and treble "boost" and -20 dB of bass and treble "cut." The frequency control point is 1600 Hz.

ACTIVE TONE CONTROL CIRCUITS

Active networks provide gain because they have amplifiers. Transistors and/or operational amplifiers are used as the active elements or amplifiers. Transistor circuits generate less distortion than operational amplifier circuits.

Active tone control circuits are Baxandall tone controls because they have four terminals, an input terminal, an output terminal, a ground terminal and a feedback-loop terminal. The feedback-loop terminal is connected between the input and the output of the active element or amplifier.

The tone control circuit of *Figure 19-3* can be used as the shunt-shunt negative-feedback network of an operational amplifier, as shown in *Figure 19-4*. The operational amplifier U1 is configured as an inverting amplifier. The operational amplifier must have a

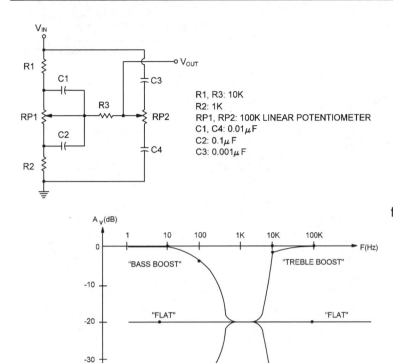

R1, R3: 10K
R2: 1K
RP1, RP2: 100K LINEAR POTENTIOMETER
C1, C4: 0.01μF
C2: 0.1μF
C3: 0.001μF

Figure 19-3.
Bass and treble
control, and its
frequency response.

large gain-bandwidth product (at least 10 MHz) or the treble section of the tone control circuit will not function. The ideal frequency response curves of the LM301 tone control circuit are shown in *Figure 19-4*. Depending upon the gain-bandwidth product (at least 10 MHz) of the operational amplifier used, the treble "boost" response begins to roll-off at 10 kHz or more. Depending upon its value, a coupling capacitors begins to roll-off the bass "boost" response at 5 Hz or less.

Most operational amplifiers require a bipolar power supply. The LM387 operational amplifier requires a single-ended power supply. An LM387 active tone control is shown in *Figure 19-5*. The operational amplifier is configured as a differential-input inverting amplifier. The non-inverting input is connected to ground via capacitor C4.

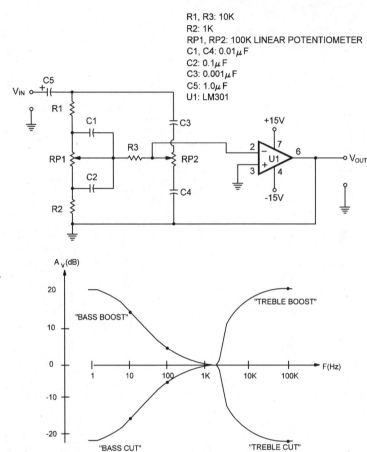

Figure 19-4.
Active tone control, and its frequency response.

There is usually an insertion loss (equal to the available "boost") associated with tone controls. The tone control is usually placed between two preamplifier stages. The LM387 tone control circuit only requires a single preamplifier because the LM387 has a high gain-bandwidth product of 15 MHz. The LM387 tone control provides 20 dB of bass and treble "boost" or "cut" when the potentiometers are fully rotated in either direction. When the potentiometers are in their mid-positions, the frequency response of the tone control circuit is "flat."

A transistor tone control is shown in *Figure 19-6*. Transistor Q1 is biased by resistors R4-R6 with a self-bias circuit. Transistor Q1 is configured as a common-emitter amplifier. The emitter resistor

R5 provides a DC negative-feedback loop to increase the stability of the quiescent operating point of Q1. Capacitor C5 bypasses R5 when an AC signal is fed to the base of Q1. The common-emitter amplifier provides a high voltage gain and a high current gain.

Transistor Q2 is configured as a common-collector (emitter-follower) amplifier. The common-collector amplifier provides a voltage gain of slightly less than unity and a high current gain. The output resistance of Q2 is low; therefore, Q2 can drive the negative-feedback network and a high impedance load, such as a preamplifier.

Transistor Q3 is configured as a constant-current source for transistor Q2. Zener diode D1 sets the reference voltage for the base of Q3 at 4.7 volts. Resistor R8 limits the current flow through zener diode D1 and through resistor R7. Resistor R7 limits the current flow through the base of Q3.

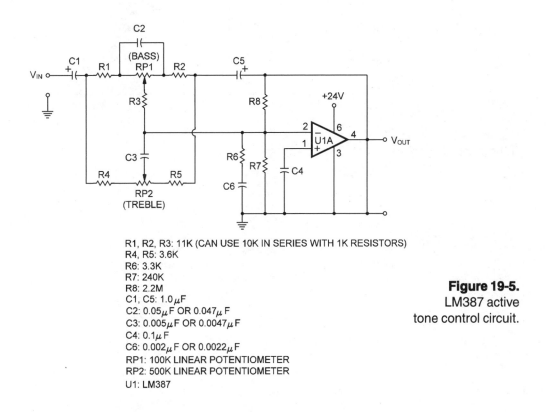

R1, R2, R3: 11K (CAN USE 10K IN SERIES WITH 1K RESISTORS)
R4, R5: 3.6K
R6: 3.3K
R7: 240K
R8: 2.2M
C1, C5: 1.0 μF
C2: 0.05 μF OR 0.047 μF
C3: 0.005 μF OR 0.0047 μF
C4: 0.1 μF
C6: 0.002 μF OR 0.0022 μF
RP1: 100K LINEAR POTENTIOMETER
RP2: 500K LINEAR POTENTIOMETER
U1: LM387

Figure 19-5.
LM387 active
tone control circuit.

The output of transistor Q2 is fed to the input of transistor Q1 via the shunt-series negative-feedback network, which consists of resistors R1-R3, potentiometers RP1 and RP2 and capacitors C1-C3.

Capacitor C4 couples the output of the bass portion of the tone control circuit to the base of transistor Q1. Capacitor C6 couples the output of transistor Q2 to the input of the shunt-series negative-feedback network.

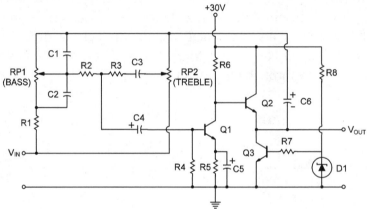

Figure 19-6.

Transistor active tone control circuit.

R1, R5: 12K
R2: 6.8K
R3: 2.2K
R4: 2.7M
R6: 120K
R7, R8: 1K
C1, C2: 47nF
C3: 2.2nF
C4, C6: 1.0μF
C5: 10μF
RP1: 100K LINEAR POTENTIOMETER
RP2: 25K LINEAR POTENTIOMETER
D1: 4.7V ZENER DIODE, 1N4732 OR 1N5230
Q1 - Q3: 2N3904

Appendix:

Problem Solutions

CHAPTER 1

1-1.   Acoustics is the physics of sound.

1-2.   A transducer is a device that converts one form of energy to another form of energy.

1-3.   A sound source, a sound receiver and a medium are required for sound.

1-4.   Sound does not travel in a vacuum. There are no particles or molecules to be vibrated by sound waves.

1-5.   Human beings can hear sounds in the range of 20 to 20,000 hertz.

1-6.   An echo is a reflected sound wave.

1-7.   Reverberation is the algebraic sum of all echoes.

1-8.   Ultrasonic sound is sound with a frequency of more than 20,000 hertz.

CHAPTER 2

2-1.   Sound waves are generated by moving particles in an elastic medium.

2-2.   The kinetic energy component of sound transmission is the movement of the particles of the medium.

2-3.   The potential energy component of sound transmission is the elastic displacement of the particles of the medium.

2-4.   The speed of sound depends on the density and on the temperature of the medium.

2-5.   Transverse wave motion occurs when the medium disturbance is perpendicular to the direction of wave propagation.

2-6.   Longitudinal wave motion occurs when the medium disturbance is parallel to the direction of wave propagation.

2-7.   Rectilinear motion is motion in a straight line.

2-8.   Simple harmonic motion is rectilinear motion whose acceleration is proportional to its displacement from a fixed point.

2-9.   Velocity is the change of position per unit time.

2-10.   Acceleration is the change in velocity per unit time.

2-11.   Wavelength is the distance between a pair of corresponding points in two consecutive waves.

2-12.   Vibration is any repetitive motion.

2-13.   Period is the time required for one complete vibration.

2-14. Frequency is the number of vibrations in one second.

2-15. Resonance occurs when the frequency of the external force equals the natural frequency of the system to which the external force is applied.

2-16. Bars, strings, membranes and circular plates are vibrators.

2-17. Standing wave is two waves of the same frequency and amplitude traveling in opposite directions through a medium.

CHAPTER 3

3-1. Spherical acoustic waves resemble plane acoustic waves at very large distances from the sound source.

3-2. The speed of sound is directly proportional to temperature. The density of the medium also affects the speed of sound.

3-3. $c = SQRT(aP/p) = 342$ m/sec.

3-4. $c = SQRT(B/p) = 1451$ m/sec.

3-5. $c = SQRT(Y/p) = 3703$ m/sec.

3-6. Three important elements of plane acoustic waves are the particle displacement, the acoustic pressure and the density change or condensation.

3-7. Acoustic pressure is the total instantaneous pressure at a point minus the static pressure.

3-8. Acoustic intensity is the average rate of flow of sound energy per unit area in the direction of wave propagation.

3-9. $I = P^2_{rms} / pc = 0.96 \times 10^{-12}$ watt/m²

3-10. Sound energy density is the energy per unit volume in a given medium.

3-11. Specific acoustic impedance is the ratio of sound pressure to the velocity of the particle.

3-12. $z = P/v = 297$ rayis

3-13. $PWL = 10 \log(W/W1) = 10 \log(50/5) = 10$ dB

3-14. Doppler effect is the apparent change in frequency when there is motion between the sound source and the observer. When the sound source and the observer approach each other, the frequency of the sound source appears to increase. When the sound source and the observer move away from each other, the frequency of the sound source appears to decrease.

3-15. Car and observer moving away from each other:
$$u = -10\,\text{m/sec, and } f' = \frac{f(c - v)}{(c - u)} = \frac{100(343 - 0)}{(343 - (-10))} = 97 \text{ Hz.}$$

Car and observer moving toward each other:
$$u = 10 \text{ m/sec, and } f' = \frac{f(c - v)}{(c - u)} = \frac{100(343 - 0)}{(343 - 10)} = 103 \text{ Hz.}$$

3-16. Acoustic intensity is the average rate of flow of sound energy per unit area.

3-17. Energy density of a spherical acoustic wave is the sum of the kinetic and potential energies per unit volume.

3-18. Specific acoustic impedance is the ratio of acoustic pressure to the velocity at any point in the wave.

3-19. Infinite baffle is a large rigid surface.

3-20.   Radiation impedance is the ratio of the force exerted by a sound radiator on the medium to the velocity of the sound radiator.

CHAPTER 4

4-1.   The human ear has a logarithmic response to sound.

4-2.   The range of the frequency response of the human ear is 20 to 20,000 hertz.

4-3.   The volume control varies the gain of an amplifier at all frequencies. The loudness control boosts the low and high frequencies as well as attenuates the medium frequencies at low volume settings.

4-4.   The bel is the logarithm to the base ten of the ratio of two powers.

4-5. $\text{bel} = \log_{10}(P2/P1) = \log_{10}(100) = 2 \text{ bel}$

4-6. $\text{dB} = 10 \log_{10}(P2/P1) = 10 \log_{10}(100) = 20 \text{ dB}$

4-7.   The threshold of pain with regard to sound intensity is 130 decibels.

CHAPTER 5

5-1.   Sound power reflection coefficient is the ratio of reflected flow of sound energy to the incident flow of sound energy.

5-2.   Sound power transmission coefficient is the ratio of transmitted sound power to the incident sound power.

5-3. $a_r = (\dfrac{p1c1 - z_2}{p1c1 + z_2})^2 = 0.5$, which yields $z_2 = 69$ rayis

5-4.   Transmission loss is the difference between sound energy striking the surface separating two spaces and the energy transmitted.

5-5. $TL = 10 \log(I_i / I_t) = 10 \log(10/5) = 3$ dB

5-6.   SWR is the ratio of the maximum amplitude to the minimum amplitude of a standing wave.

5-7. $SWR = \dfrac{Pi + P_r}{P_i - P_r}$,   from which

$\dfrac{Pr}{Pi} = \dfrac{SWR - 1}{SWR + 1}$

5-8.   Snell's law states that the angle of incidence equals the angle of reflection.

5-9.   Critical angle is the minimum angle at which all incident sound waves are reflected.

5-10. Sound waves whose wavelengths are about the same as the dimension of the obstacle are diffracted.

5-11. Sound waves whose wavelengths are large compared to the dimension of the obstacle are scattered.

5-12. Destructive interference occurs when two sound waves of the same frequency and amplitude meet out of phase.

5-13. Constructive interference occurs when two sound waves of the same frequency and amplitude meet in phase.

CHAPTER 6

6-1.   The loudspeaker and the microphone are transducers. A speaker converts electrical energy into acoustical energy. A microphone converts acoustical energy into electrical energy.

6-2.   Moving coil, velocity ribbon and magnetostriction microphones are constant-velocity microphones. Carbon, condenser and crystal microphones are constant-amplitude microphones.

6-3. $M = EhA/Rs = 0.02355$ volt/nt-m^2
and $20\log(M/10) = -52.56$ dB

6-4.   The resonance of a microphone can be controlled by damping the diaphragm and/or by making the microphone housing thick and heavy.

6-5.   The permanent magnet and the voice coil make up the dynamic speaker's driver.

6-6.   No, because the impedance is the AC resistance.

6-7.   Electrodynamic speakers require an external power supply to energize its magnetic field.

6-8.   The condenser or electrostatic speaker has a high impedance.

6-9.   The electrostatic speaker is not suitable for use as a woofer because its electrodes have to be closely spaced.

6-10.   The crystal speaker is not useful as a woofer because it produces low power and it has a limited low frequency response.

6-11. $W = (BL)^2 RE^2/Z_m^2 Z_i^2 = 1.86W$

6-12. A woofer is a speaker designed to respond to low frequencies. A tweeter is a speaker designed to respond to high frequencies.

6-13. Free-air resonance is the frequency at which a speaker cone resonates.

6-14. Woofers require large magnets to move the speaker cone large distances and to reproduce high volume low frequency sounds.

6-15. A speaker's suspension determines its compliance.

6-16. Two electrical analogies for acoustical systems are the voltage-pressure analogy and the current-pressure analogy.

CHAPTER 7

7-1. Noise is non-periodic vibration. Music is periodic vibration.

7-2. Ultrasonic noise occurs above the audible band of frequencies. Infrasonic noise occurs below the audible band of frequencies.

7-3. Loud, low pitch sounds cause muscles to tighten human eardrums.

7-4. Loudness cannot be measured because it is a subjective interpretation.

7-5. No, the human ear is most sensitive to sounds in the 1000 to 5000 hertz range at low sound intensities.

7-6. $ISL = 10\log(I/I_o \, dF) = 37 \, dB$

$Il = ISL + 10\log(dF) = 37 + 30 = 67 \, dB$

7-7.   White noise has a constant spectrum level over the band of audible frequencies. Pink noise has equal energy per octave over the same band of audible frequencies.

7-8.   Musical sounds differ from each other in pitch, timbre, loudness and intensity.

7-9.   The timbre of a sound is determined by the harmonic frequencies present and the intensities of the harmonic frequencies present.

7-10. A major triad is three frequencies with the ratio 4:5:6.

7-11. A major chord is four frequencies with the ratio 4:5:6:8.

7-12. A keynote is the first note of any octave.

7-13. A beat occurs when two sound waves of slightly different frequencies alternately reinforce and cancel each other.

7-14. Three types of musical instruments are wind instruments, string instruments and percussion instruments.

7-15. The vocal cords are located in the larynx.

7-16. Vowel sounds are musical sounds. Consonant sound are noise sounds.

7-17. Laryngitis causes vocal cords to thicken. Thicker vocal cords vibrate more slowly, lowering the pitch of the voice.

7-18. Your voice does not sound the same to you as it does to others. It travels to others through air. It travels to you through the air and through the medium of your jaw bone.

7-19. Five elements of good speech are breathing, pitch, volume, velocity and timbre.

7-20. Some speech qualities are twang, nasal, lisp and drawl.

7-21.   Phonetics is the study of speech.

7-22.   A phoneme is the smallest unit of speech.

CHAPTER 8

8-1. $SL = 10\log(I/I1) = 10\log(10^{-6}/10^{-7}) = 10$ dB

8-2.   Conduction deafness is hearing loss due to an abnormality or obstruction of the middle ear. Nerve deafness is hearing loss due to a nerve defect or damage.

8-3.   Tone deafness is the inability to discriminate small differences in pitch.

8-4.   Binaural audition is the ability to locate the direction of a sound source.

8-5.   Sounds of frequencies less than seven hundred hertz are located by the phase difference between the human ears.

8-6.   Sounds of frequencies in the range of 700 to 5000 hertz are located by head movement.

8-7.   Sounds of frequencies above 5000 hertz are located by the pinnae of the human ears.

8-8.   The human ear is about the size of a hazelnut.

8-9.   The ear converts sound waves into nerve impulses. It also helps maintain the equilibrium or balance of a moving person.

8-10.   The cochlea converts sound waves into nerve impulses.

8-11. The basilar membrane detects the pitch and amplitude of the sound and it transmits this information to the cochlear nerve fibers.

8-12. Sound perception and interpretation occur in the brain.

CHAPTER 9

9-1. Reverberation is the persistence of sound in a room.

9-2. Absorbing materials are used to control reverberation.

9-3. One second is the ideal reverberation time for speech. At least two seconds is the ideal reverberation time for music.

9-4. Direct sound arrives at the listener's ear directly from the source. Reverberant sound arrives at the listener after one or more reflections from either the surfaces of the concert hall or from objects inside the concert hall.

9-5. Reverberation time is time in seconds for sound pressure to decrease by sixty decibels of its original value after the sound source is turned off.

9-6. $T = 0.049V/a = 0.049 \times 500,000/10,000 = 2.45$ seconds

9-7. If the reverberation time of a room is too short, the sound intensity is not even in all areas of the room. If the reverberation time of a room is too long, echoes are present.

9-8. A reverberation chamber is a room with surfaces having minimal sound absorption. The surfaces are reflective.

9-9. $I(0.25) = 91 \times 10^{-8}$ watt/m^2.

9-10.   Airborne noise enters buildings through holes, cracks, air intake vents, air exhaust vents and poorly fitting doors and windows.

9-11.   Transmission loss is the difference in decibels between the sound energy striking the surface separating two spaces and the sound energy transmitted.

9-12.   Vibrating elastic bodies cause structural-borne noise.

9-13.   Structural-borne noise should be suppressed at its source because it contains a lot of energy. It is easy and economical to suppress structural-borne noise at its source.

9-14.   Sound energy that is absorbed is converted to heat and to kinetic energy.

9-15.   Sound absorption coefficient is the decimal fraction of perfect absorption that it possesses.

9-16. $RF = TL + 10 \log(a/S) = 20 - 10 = 10$ dB

9-17.   An anechoic chamber has walls covered in highly absorptive material.

9-18.   Good architectural acoustics requires smooth growth of sound and smooth decay of sound.

9-19.   Axial sound waves decay slowly. Oblique sound waves decay quickly.

9-20.   A reflecting surface must be at least 55 feet away from the sound source to generate echoes.

9-21.   Room flutter is multiple echoes.

9-22.   Sound focusing is the concentrating of sound at a point.

9-23.   A dead spot is a region of sound deficiency.

CHAPTER 10

10-1.   Water temperature, pressure gradients, marine organisms, air bubbles and salt content can affect the transmission of sound waves underwater.

10-2.   Divergence, absorption and irreversible attenuation cause sound transmission losses in salt water.

10-3.   Sound waves are refracted downward in an arc underwater because temperature varies linearly with the depth of the water.

10-4.   Sound waves are refracted upward in an arc underwater because pressure increases linearly with the depth of the water.

10-5.   Ambient noises in the sea are caused by wind, rain and the state of agitation of the sea.

10-6.   A hydrophone is a transducer that converts sound waves in water into electrical energy.

10-7.   Hydrophone directivity is an indication of the fraction of the total signal it converts into electrical energy according to its sensitivity pattern.

10-8.   Underwater sound projector is a transducer that converts electrical energy into acoustical energy in the water.

10-9.   CN = 3.33

10-10. Sonar is used to detect oil deposits, enemy ships, enemy submarines and other underwater objects.

10-11. Active sonar transmits sounds and listens for echoes that are reflected by the target. Passive sonar does not transmit sounds. Passive sonar listens for sounds produced by the target.

CHAPTER 11

11-1. Ultrasonics is the physics of sound waves having frequencies above the limits of human hearing.

11-2. The thermal lattice vibration beyond which a material can not follow the input sound wave limits the upper frequency for the propagation of ultrasonic sound waves.

11-3. Rayleigh waves are used to detect flaws and cracks on or near the surface of a test object.

11-4. Lamb waves are used to locate areas not bonded in laminated structures, to locate radial cracks in tubing and for quality control of plate and sheet stock.

11-5. Gas-driven, liquid-driven and electromechanical transducers are the three types of ultrasonic transducers.

11-6. From chapter three (plane and spherical acoustic waves):

$$c = SQRT(Y/p) = SQRT(7.9 \times 10^{10}/2.65 \times 10^3) = 5460 \text{ m/sec}$$

$$F = c/2t = 5460/0.002 = 2.73 \text{ MHz.}$$

11-7. $WL = 2L = 2 \times 0.2 = 0.4 \text{ meters.}$

$F = c/WL = 4900/0.4 = 12.25 \text{ kHz}$

11-8. Gas absorbs ultrasonic energy by heat conduction and due to the viscosity of the gas.

11-9.   Ultrasonics can be propagated much further in water than in gases or solids. Higher frequency ultrasonics can be propagated in water than in gases or solids. Water does not attenuate or absorb ultrasound very much.

11-10. An electrical signal converted into an ultrasound wave travels through the copper wire at 3700 m/sec. At the end of the copper wire, the ultrasound wave is converted back into the original electrical signal.

   $L = ct = 3700 \times 10^{-9} = 3.7 \times 10^{-6}$ meters.

Glossary

Absorb: take in sound waves by molecular action.

Acceleration: change of velocity per unit time.

AC coupling: coupling that allows only time varying signals to pass through it.

Acoustic feedback: squealing sound caused by feeding the output signal of an audio circuit back to the input of the audio circuit in phase with the input signal.

Acoustic fiberglass: material used to absorb sound waves inside speaker enclosures.

Acoustic lens: horn designed to control the directional spread of sound.

Acoustic suspension speaker: speaker designed for use in a sealed enclosure.

Acoustic intensity: average rate of flow of sound energy per unit area in the direction of wave propagation.

Acoustical compliance: the ratio of volume displacement to acoustic pressure.

Acoustical inertance: the ratio of acoustic pressure to the rate of change of the volume velocity.

Acoustical resistance: the ratio of acoustic pressure to volume velocity.

Acoustics: physics of sound.

Acoustic shadow: obstacle to sound waves.

Adiabatic: occurring without heat loss or heat gain.

Air suspension speaker: an acoustic suspension speaker.

Alternating current: time varying electrical current.

Ampere: unit of electrical current in coulombs per second.

Amplifier: a device used to increase the amplitude of a signal.

Amplitude: relative strength of a signal.

Anechoic chamber: room with totally sound-absorbent walls.

Attenuation: reduction of a signal.

Audio frequency: the frequency range of human hearing, twenty hertz to twenty thousand hertz.

Axial: not deviating from the perpendicular.

Baffle: piece of wood inside an enclosure used to block or direct sound waves.

Balance: relative signal strength of the two channels of a stereo system.

Bandpass filter: electric circuit passing middle frequencies of a signal.

Basket: metal frame of a speaker.

Bass: low frequency sounds, twenty hertz to one thousand hertz.

Bass reflex enclosure: a ported reflex speaker enclosure.

Beam width: angle at which sound intensity drops to one-half its value at the axial direction of the source.

Beat: two sounds of slightly different frequencies alternately reinforcing and canceling each other.

Bel: the logarithm to the base ten of the ratio of two powers.

Binaural audition: ability to locate the direction of a sound source.

Bobbin: cylinder around which is wound the speaker voice coil. The bobbin is mechanically connected to the speaker cone.

Capacitor: electronic component used to store electrical charge between its two parallel plates. The plates are separated by a dielectric or insulating material.

Ceruminous glands: wax glands.

Channel: left or right signal of a stereo system.

Circuit: complete path that allows an electric current to flow from one terminal of a voltage source to the other terminal of a voltage source.

Cleats: strips of wood used inside an enclosure to reinforce corners or to provide a mounting surface for the front and rear panels.

Clipping: distortion caused by clipping the peaks of a signal. Clipping is usually caused by an overdriven amplifier.

CMRR: common-mode rejection ratio is the ability of an operational amplifier to cancel out, within the device, common signals fed to the inverting and non-inverting inputs.

Coaxial speaker: speaker with two voice coils and cones and it is used to reproduce sounds in two segments of the sound spectrum.

Compliance: relative stiffness of a speaker suspension.

Condensation: compression of longitudinal waves or density change.

Conduction deafness: hearing loss due to abnormality or obstruction of the middle ear.

Cone: cone-shaped diaphragm of a speaker attached to the voice coil. Produces sound waves when it vibrates.

Continuants: prolonged holding to an initial phoneme.

Critical angle: minimum angle where sound waves are reflected.

Crossover network: electronic circuit that splits audio frequencies into different bands suitable for individual speakers.

Current: flow of electrons. Current is measured in amperes.

Cutoff frequency of a horn: frequency below which there is no propagation of sound waves inside the horn.

Damping: controlling the movement of a speaker cone by the speaker suspension and/or the pressure inside the speaker enclosure.

Dead spot: region of sound deficiency.

Deafness: hearing loss measured in decibels.

Decibel: lowest sound intensity that the human ear can detect.

Diffract: deflection of sound waves by an obstacle or when passing through an opening.

Diffraction horn: narrow horn that expands uniformly in the vertical direction but it is unflared in the horizontal direction.

Diffuse echo: scattering of sound waves by several obstacles.

Direct current: current flowing in one direction.

Directional efficiency: of a microphone is the ratio of energy output due to simultaneous sounds at all angles to energy output which would be obtained from an omnidirectional microphone with the same axial sensitivity.

Directivity: variation of the microphone output with different angles of incidence. Directed response characteristic.

Directivity ratio: ratio of intensity at any point on the axis of the sound source to the intensity that would be produced at the same point by a simple source of equal strength.

Direct sound: sound arriving at listener's ears directly from the sound source.

Dispersion: spreading of sound waves as they leave the speaker.

Dissipative losses: power losses of an inductor due to its wire windings.

Distortion: any unwanted change in an electrical signal.

Dome tweeter: high frequency speaker with a dome-shape diaphragm. Provides better dispersion than standard cone speakers.

Doppler effect: apparent change in frequency when there is motion between the sound source and the observer.

Drawl: speak or pronounce slowly.

Driver: the magnet and voice coil of a speaker.

Drone: passive radiator or speaker with a cone but no driver components. Cone vibrates in response to air pressure changes in the enclosure. The drone increases bass output without an increase in electrical power.

Ducted port: a ported reflex speaker enclosure.

Dynamic range: range of sound levels which a system can reproduce without introducing distortion. It is usually expressed in decibels.

Echo: reflected sound wave.

Energy density: sum of kinetic and potential energies per unit volume.

Equalization: adjustment of frequency response to tailor the sound to match personal preferences, room acoustics and/or speaker enclosure design.

Equalizer: an electronic circuit that alters a signal to provide a desired frequency response.

Equilibrium: balance.

Extraneous: coming from the outside.

Farad: unit of capacitance in coulomb per volt.

Filter: electric circuit that passes or blocks signals of certain frequencies.

Flat response: reproduction of an audio signal within plus or minus one decibel.

Free-air resonance: natural resonant frequency of a woofer when operating outside an enclosure.

Frequency: number of vibrations in one second.

Frequency response: range of frequencies accurately reproduced by a system.

Full range: speaker designed to reproduce most or all of the sound spectrum.

Fundamental tone: lowest frequency component of a signal.

Glides: continuous movement of the vocal mechanism from one phoneme to another phoneme.

Grille cloth: fabric used to cover speakers mounted in an enclosure.

Harmonic distortion: harmonic tones artificially added by an electrical circuit or speaker. It is expressed as a percentage of the original signal.

Harmonic echo: differential scattering of a complex sound of different frequencies.

Harmonic tone: multiples of the fundamental tone, usually generated by the interaction of signal waveforms.

Hearing loss: decibel difference between subject's threshold of hearing and that for a person with normal hearing at a given frequency.

Hertz: unit of frequency equalling one cycle per second. Named in honor of German physicist H.R. Hertz.

High pass filter: electric circuit passing only high frequencies.

Hiss: noise that sounds like an air leak.

Horn: speaker or enclosure design using its own funnel-shape to disperse sound waves.

Hum: audio noise that has a steady low frequency pitch.

Hydrophone directivity: indication of the fraction of the total signal it converts into electrical energy according to its sensitivity pattern.

Hydrophone: transducer that converts sound waves in water into electrical energy.

Hygroscopic: absorbing moisture from the air.

Impedance: opposition of an electric circuit or speaker to an alternating current.

Inductance: the ability of a coil to store energy in a magnetic field that surrounds it. Inductance resists the flow of an alternating current.

Infinite baffle: large rigid surface.

Integrated amplifier: preamplifier and power amplifier combined in one unit.

Intensity: power transmitted per unit area.

Intensity spectrum level: at any frequency is the intensity level of a given noise contained within a band of frequencies one hertz wide centered on that frequency.

Interfere: to impede sound waves.

Isotropic: material whose properties are the same in all directions.

Keynote: first note of any octave.

Lisp: speech defect where sibilants "S" and "Z" are articulated like "TH" in thank and "TH" in this.

Longitudinal wave motion: wave motion where medium disturbance is parallel to the direction of propagation of the wave.

Loudspeaker: electroacoustic transducer converting electrical energy into acoustical energy.

Low pass filter: electric circuit passing only low frequencies.

L-pad: potentiometer used to maintain a constant impedance at its input while varying the signal level at its output.

Magnetostrictive rod: rod whose length changes when it is exposed to a varying magnetic field.

Major chord: four frequencies with the ratio 4:5:6:8.

Major triad: three frequencies with the ratio 4:5:6.

Mel: acoustic unit to describe the pitch of a sound.

Microphone: electroacoustic transducer converting acoustic energy into electrical energy.

Mid-range: speaker designed to reproduce middle frequencies of the sound spectrum, that is, the range of one thousand hertz to four thousand hertz.

Mounting flange: outer edge of speaker frame with predrilled holes to accept screws or bolts which are used to secure the speaker to the enclosure.

Multicellular array: group of horns driven by a common source. Each horn acts like a separate sound radiator.

Music: periodic vibration.

Nasal: produced with voice passing through the nose.

Nerve deafness: hearing loss due to a nerve defect.

Noise: any unwanted sound. Non-periodic vibration.

Nominal power rating: continuous or RMS power rating.

Noy: acoustic unit to rank annoyance of noises to human ears.

Oblique: deviating from the perpendicular.

Octave: interval between two frequencies with a ratio of 2:1.

Ohm: unit of electrical resistance or impedance.

Ohm's law: law of electric circuits that states the current flow I (in amperes) through an electric circuit is equal to the voltage source V (in volts) divided by the resistance R (in ohms) of the circuit. I=V/R

Ossicle: bone.

Passive radiator: speaker with a cone but without driver components. It vibrates due to changes in pressure inside the speaker enclosure. It increases the bass output without an increase in electrical power. Also called a drone.

Peak: maximum amplitude of a voltage or current.

Period: time required for one complete vibration.

Phon: acoustic unit to measure overall loudness level of a noise.

Phoneme: smallest unit of speech.

Phonetics: study of speech sounds.

Piezoelectric: crystal that vibrates when a voltage is applied to its surfaces.

Pink noise: noise with equal energy per octave from twenty hertz to twenty thousand hertz.

Pinnae: flaps of the ears.

Plenum: space completely occupied by matter.

Polarity: orientation of an electric or magnetic field. The direction of movement of a speaker cone depends on the polarity of the audio signal applied to the speaker terminals.

Ported reflex enclosure: speaker enclosure using a tuned duct or port to improve efficiency at low frequencies.

Pressure spectrum level: sound pressure level contained within a band of frequencies one hertz wide.

PSRR: or power supply rejection ratio is the ability of an operational amplifier to prevent power supply fluctuations from showing up in the output signal.

Radiation efficiency of a horn: ratio of actual acoustic power radiated out of the horn to the acoustic power radiated by the same diaphragm which moves at the same velocity into a cylindrical tube of infinite length and having the same cross-sectional area as the throat of the horn. Also called the transmission coefficient.

Radiation impedance: ratio of the force exerted by a sound radiator on the medium to the velocity of the sound radiator.

Rarefaction: expansion of longitudinal waves.

Rectilinear motion: motion in a straight line.

Reflect: to turn or throw back a sound wave.

Refraction: change of direction of a sound wave when passing from one medium to another medium.

Resonance: frequency of the external force equals the natural frequency of the system.

Reverberant sound: sound reaching listener after one or more reflections from either the surfaces of a hall of from the objects inside the hall.

Reverberation: algebraic sum of all echoes. Persistence of sound.

Reverberation chamber: room with surfaces having minimal sound absorption.

Reverberation time: time in seconds for sound pressure to decrease by sixty decibels of its original value after the sound source is turned off.

Room flutter: multiple echoes in the room.

RMS: acronym for root mean square. The RMS value of an alternating current produces the same heating effect in an electric circuit as the same value (of the RMS of an alternating current) of direct current.

Scatter: sound waves separate and go in different directions.

Sebaceous glands: oil glands.

Sensation level: of a tone is the amount it exceeds the threshold of hearing, in decibels.

Sensitivity: voltage output of a microphone for a sound pressure input of one microbar. Open-circuit voltage microphone response.

Sibilant: consonants produced by the frictional passage of breath through a narrow opening in the front part of the mouth.

Signal: desired part of an electric signal.

Signal-to-noise: the ratio in decibels between the signal and the noise. Also referred to as S/N.

Simple harmonic motion: rectilinear motion whose acceleration is proportional to its displacement from a fixed point.

Sinewave: waveform of an alternating signal. It varies about a zero point to a positive and negative value. The positive and negative values are equal.

Slew rate: time required for an amplifier's output to respond to an input signal.

Sone: acoustic unit to measure loudness of sound.

Sonic wave: high amplitude ultrasonic wave.

Sound absorption coefficient: of a material is the decimal fraction of perfect absorption that it possesses.

Sound articulation: percentage of total number of speech sounds correctly identified.

Sound energy density: energy per unit volume in a given medium.

Sound focusing: concentration of sound at a point.

Sound level meter: used to measure sound energy in decibels.

Sound power reflection coefficient: (1) ratio of reflected flow of sound energy to the incident flow of sound energy; (2) ratio of reflected sound power to the incident sound power.

Sound pressure level: loudness of sound produced by a sound source such as a speaker or stereo system.

Sound transmission loss: losses in sea water due to divergence, absorption and irreversible attenuation.

Specific acoustic impedance: ratio of sound pressure to the velocity at any point in the wave.

Speech interference level: average of readings in three octave frequency bands, in decibels.

Spider: flexible fabric that supports the bobbin, voice coil and inside part of the cone within the speaker frame.

Standing wave: two waves of the same frequency and amplitude travelling in opposite directions through a medium.

Standing wave ratio: ratio of acoustic pressure at an antinode to acoustic pressure at a node. Ratio of maximum amplitude to minimum amplitude in a standing wave.

Stapedius: small muscle of the middle ear.

Stops: blocking and unblocking of the air flow through the larynx.

Surround: outer suspension of a speaker cone. The surround connects the outside part of the cone to the speaker frame.

Surround sound: the reproduction of the spacious acoustics of a live performance in a residential listening room.

Suspension: the surround of a speaker.

Syllable articulation: percentage of total number of syllables correctly identified.

Tensor tympani: small muscle of the middle ear.

Three-way: speaker system composed of three speakers, each with a different frequency range of operation.

Timbre: sound quality.

Tone deafness: inability to discriminate small differences in pitch.

Total harmonic distortion: percentage, in relation to a pure input signal, of harmonically derived frequencies introduced by a system.

Tourmaline: mineral occurring in various colors.

Transducer: device that converts one form of energy into another form of energy.

Transmission coefficient of a horn: see radiation efficiency of a horn.

Transmission loss: difference in decibels between sound energy striking the surface separating two spaces and the energy transmitted.

Transverse wave motion: motion where medium disturbance is perpendicular to the direction of propagation of the wave.

Treble: upper end of the audio spectrum.

Triaxial driver: speaker with three voice coils and cones, used to reproduce sounds in three segments of the sound spectrum.

Twang: to speak nasally.

Tweeter: speaker designed to reproduce high frequency sounds above four thousand hertz.

Two-way: speaker system composed of two speakers, each with a different frequency range of operation.

Ultrasonics: physics of sound waves of frequencies above the limits of human hearing.

Underwater sound projector: transducer that converts electrical energy into acoustical energy in the water.

Unity-gain bandwidth: the highest frequency that an amplifier will pass at a gain of unity without attenuation in the amplification process.

Velocity: change of position per unit time.

Vibration: any repetitive motion.

Voice coil: wire wound around the speaker bobbin. The bobbin is mechanically connected to the speaker cone. The bobbin causes the cone to vibrate in response to the audio signal in the voice coil.

Watt: unit of electrical power.

Wavelength: distance between a pair of corresponding points in two consecutive waves.

Wave train: succession of waves caused by a continuously vibrating source.

White noise: noise with a constant spectrum level over the band of audible frequencies.

Whizzer: small supplementary cone attached to the center of the speaker's main cone for the purpose of increasing the high frequency response.

Woofer: speaker designed to reproduce low frequency sounds less than one thousand hertz.

Bibliography

ABC's Of The Human Body, Reader's Digest Press, New York, 1987.

The ARRL Handbook, 70th Edition, The American Radio League, Connecticut, 1993.

Badmaieff, A. and Davis, D. *How To Build Speaker Enclosures*, Howard W. Sams & Co., Inc., Indiana, 1968.

Basic Electronics, Bureau Of Naval Personnel, Washington, D.C., 1968.

Beer, F.P. and Johnson Jr., E.R. *Mechanics for Engineers, Dynamics, Third Edition*, McGraw-Hill Inc, New York, 1976.

Bose, A.G. "Sound Recording and Reproduction", *Technology Review*, June 1973 and July/August 1973.

COS/MOS Integrated Circuits, RCA Corporation, New Jersey, 1974.

Dienhart, C.M. *Basic Human Anatomy And Physiology, Third Edition*, W.B. Saunders Company, Pennsylvania, 1979.

Eubank, H.L., Ramsay, J.M. and Rickard, L.A. *Basic Physics For Secondary School*, The Macmillan Company Of Canada, Ontario, 1957.

Gerzon, M. "Surround-Sound Psychoacoustics", *Wireless World*, December 1974.

Lenk, J.D. *Handbook Of Practical Electronic Tests And Measurements*, Prentice-Hall Inc., New Jersey, 1969.

Linear Applications, National Semiconductor Corporation, California, 1973.

Linear Integrated Circuits, National Semiconductor Corporation, California, 1973.

Lurch, E.N. *Fundamentals of Electronics, Second Edition*, John Wiley & Sons Inc., New York, 1971.

Marston, R.M. "Audio Pre Amp IC's", *Radio-Electronics*, Gernsback Publications Inc., New York, February 1990.

McComb, G. *Building Speaker Systems*, Master Publishing Inc., Texas, 1988.

Melen, R. and Garland, H. *Understanding IC Operational Amplifiers*, Howard W. Sams & Co., Inc., Indiana, 1973.

Prochnow, D. *Chip Talk: Projects In Speech Synthesis*, Tab Book Inc., Pennsylvania, 1987.

Ratcliff, J.D. *Your Body & how it works*, Reader's Digest Press, New York, 1978.

Schultz, R.W. and Lagemann, R.T. *Physics For The Space Age*, J.B. Lippincott Company, Illinois, 1961.

Seto, W.W. *Acoustics*, McGraw-Hill Book Company, New York, 1971.

Singmin, A. "Bass and Treble Booster Controls", *Electronics Hobbyists Handbook* (Popular-Electronics), Gernsback Publications Inc., New York, 1993.

Index

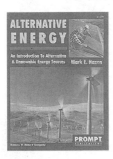

Alternative Energy
by Mark Hazen

Home Security Projects
by Robert Gaffigan

This book is designed to introduce readers to the many different forms of energy mankind has learned to use. Generally, energy sources are harnessed for producing electricity. This process relies on transducers to transform energy from one form into another. *Alternative Energy* will not only address transducers and the five most common sources of energy that can be converted to electricity, it will also explore solar energy, the harnessing of the wind for energy, geothermal energy, and nuclear energy.

This book is designed to be an introduction to energy and alternate sources of electricity. Each of the nine chapters are followed by questions to test comprehension, making it ideal for students and teachers alike. In addition, listings of World Wide Web sites are included so that readers can learn more about alternative energy and the organizations devoted to it. Author Mark Hazen is currently a professor of electronics at Brevard Community College.

Professional Reference
285 pages • paperback • 7-3/8 x 9-1/4"
ISBN: 0-7906-1079-5 • Sams: 61079
$24.95

Home Security Projects presents the reader with many projects about home security, safety and nuisance elimination that can easily be built in the reader's own home for less than it would cost to buy these items ready-made. Readers will be able to construct devices that will allow them to protect family members and electrical appliances from mishaps and accidents in the home, and protect their homes and belongings from theft and vandalism.

This book shows the reader how to construct the many useful projects, including a portable CO detector, trailer hitch alignment device, antenna saver, pool alarm, dog bark inhibitor, and an early warning alarm system. These projects are relatively easy to make and the intent of *Home Security Projects* is to provide enough information to allow you to customize them.

Projects
256 pages • paperback • 6 x 9"
ISBN: 0-7906-1113-9 • Sams: 61113
$24.95

CALL 1-800-428-7267 TODAY FOR THE NAME OF YOUR NEAREST PROMPT PUBLICATIONS DISTRIBUTOR
Prices subject to change.

PROMPT®
PUBLICATIONS

Desktop Digital Video
by Ron Grebler

The Video Hacker's Handbook
by Carl Bergquist

Desktop Digital Video is for those people who have a good understanding of personal computers and want to learn how video (and digital video) fits into the bigger picture. This book will introduce you to the essentials of video engineering, and to the intricacies and intimacies of digital technology. It examines the hardware involved, then explores the variety of different software applications and how to utilize them practically. Best of all, *Desktop Digital Video* will guide you through the development of your own customized digital video system. Topics covered include the video signal, digital video theory, digital video editing programs, hardware, digital video software and much more.

Geared toward electronic hobbyists and technicians interested in experimenting with the multiple facets of video technology, *The Video Hacker's Handbook* features projects never seen before in book form. Video theory and project information is presented in a practical and easy-to-understand fashion, allowing you to not only learn how video technology came to be so important in today's world, but also how to incorporate this knowledge into projects of your own design. In addition to the hands-on construction projects, the text covers existing video devices useful in this area of technology plus a little history surrounding television and video relay systems.

Video Technology
225 pages • paperback • 7-3/8 x 9-1/4"
ISBN: 0-7906-1095-7 • Sams: 61095
$34.95

Video Technology
336 pages • paperback • 7-3/8 x 9-1/4"
ISBN: 0-7906-1126-0 • Sams: 61126
$29.95

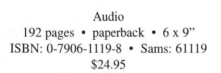